The Patrick Moore Practical Astronomy Series

For further volumes:
http://www.springer.com/series/3192

Getting Started in Radio Astronomy

Beginner Projects for the Amateur

Steven Arnold

Steven Arnold
Mansfield, UK

ISSN 1431-9756
ISBN 978-1-4614-8156-0   ISBN 978-1-4614-8157-7 (eBook)
DOI 10.1007/978-1-4614-8157-7
Springer New York Heidelberg Dordrecht London

Library of Congress Control Number: 2013946929

© Springer Science+Business Media New York 2014
This work is subject to copyright. All rights are reserved by the Publisher, whether the whole or part of the material is concerned, specifically the rights of translation, reprinting, reuse of illustrations, recitation, broadcasting, reproduction on microfilms or in any other physical way, and transmission or information storage and retrieval, electronic adaptation, computer software, or by similar or dissimilar methodology now known or hereafter developed. Exempted from this legal reservation are brief excerpts in connection with reviews or scholarly analysis or material supplied specifically for the purpose of being entered and executed on a computer system, for exclusive use by the purchaser of the work. Duplication of this publication or parts thereof is permitted only under the provisions of the Copyright Law of the Publisher's location, in its current version, and permission for use must always be obtained from Springer. Permissions for use may be obtained through RightsLink at the Copyright Clearance Center. Violations are liable to prosecution under the respective Copyright Law.
The use of general descriptive names, registered names, trademarks, service marks, etc. in this publication does not imply, even in the absence of a specific statement, that such names are exempt from the relevant protective laws and regulations and therefore free for general use.
While the advice and information in this book are believed to be true and accurate at the date of publication, neither the authors nor the editors nor the publisher can accept any legal responsibility for any errors or omissions that may be made. The publisher makes no warranty, express or implied, with respect to the material contained herein.

Cover illustration: Cover photo of the CSIRO Parkes Radio Telescope taken by Amanda Slater and used via creative commons license.

Printed on acid-free paper

Springer is part of Springer Science+Business Media (www.springer.com)

Dedication

For Marjorie,
I can't imagine life without you or Sweep.

With special thanks to;

Shlomo, Lorraine and Josh,
for their patience and proof reading
skills and to Linda T for your encouragement.

Also to the memory of three truly inspirational
people that sadly died last year (2012).

Sir Bernard Lovell; Radio astronomer at Jodrell Bank.
Neil Armstrong; First man on the Moon.
Sir Patrick Moore; Astronomer and true gentleman.

About the Author

Steven Arnold is a certified mechanical engineer and long-time amateur radio astronomer. He is a member of the Society for Popular Astronomy and has recently contributed radio astronomy articles about the NASA Radio Jove receiver kit and also wrote about the Stratospheric Observatory for Infrared Astronomy (SOFIA). He wrote and recorded a podcast about the NASA Radio Jove project, which can be heard on www.365daysofastronomy.org. Steve is a member of a number of astronomical groups and organizations. He has his own small permanent observatory and specializes in solar system deep-sky imaging, spectroscopy, and radio astronomy.

Contents

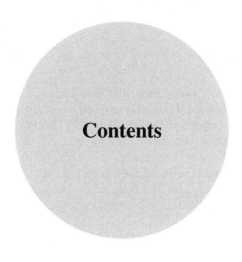

1	**The History of Radio Astronomy**		1
	1.1 The Early Years		1
	1.2 Wartime Developments		7
	1.3 Increasing Resolution		11
	1.4 Planetary Signals		16
	1.5 Sputnik		17
	1.6 The Search for Intelligent Life		19
	1.7 The Apollo Mission		20
	1.8 Interferometry and Quasars		21
	1.9 Observation of Mercury		23
	1.10 The 'Big Bang' Theory and CMB		24
	1.11 Neutron Stars and Pulsars		26
	1.12 Arecibo Transmission and 'Wow!' Signal		28
2	**The Basic Physics of Electricity and Magnetism**		33
	2.1 Building Blocks		33
	2.2 Magnetism and Wave Properties		35
	2.3 The Electromagnetic Spectrum		39
	2.4 Radio Frequencies and Characteristics		42
3	**Atmosphere, Magnetic Fields and Magnetospheres**		49
	3.1 The Goldilocks Zone		49
	3.2 The Ionosphere		51
	3.3 Scintillation		54
	3.4 Planetary Magnetic Fields and Magnetospheres		55

ix

4	**Electrical Components**	61
4.1	A Brief Review of Electrical Safety	61
4.2	Soldering	62
4.3	Multimeters	66
4.4	Electrical Component Identification	68
4.5	Headphones	77
5	**The Stanford Solar Center, SuperSID Monitor**	81
5.1	Space Weather and Its Dangers to Earth	84
5.2	A Basic Description of How the SuperSID Monitor Works	85
5.3	How to Make an Antenna	88
5.4	Placing the SuperSID Antenna	94
5.5	Computer Sound Cards and Software	95
5.6	Connecting Everything Together	98
5.7	Interpreting the Data	102
5.8	X-ray Classification	105
5.9	Resources for More Information	106
6	**The NASA INSPIRE Project**	109
6.1	What Is the INSPIRE Project?	109
6.2	A Guide to Building an INSPIRE Receiver	111
6.3	Manmade VLF Radio Emissions and How to Choose an Observing Site	119
6.4	Natural Radio Emissions	122
6.5	Radio Emissions from Beyond the Grave?	125
6.6	Recording VLF	126
6.7	Analyzing Software and Where to Get It	129
6.8	References for More Information	133
7	**The NASA Radio Jove Project**	135
7.1	About Radio Jove	135
7.2	A Guide to Building the Radio Jove Receiver	137
7.3	Building the Radio Jove Antennas	144
7.4	Antenna Configurations	150
7.5	Software	154
7.6	Calibration	158
7.7	Radio Emissions from the Sun	160
7.8	Radio Emissions from the Planet Jupiter	164
7.9	References for More Information	166
8	**Radio Detection of Meteors**	169
8.1	Meteoroid, Meteor, Meteorite and Micrometeorites	169
8.2	What Is Radio Detection of Meteors?	171
8.3	When to Listen	173
8.4	What Equipment to Use	175
8.5	Software for Recording and for Analyzing	180
8.6	References for More Information	183

9	**Conclusions**	187
	9.1 The INSPIRE Receiver	188
	9.2 Radio Meteor Detection	188
	9.3 The SuperSID Monitor	188
	9.4 The Radio Jove Project	189
Glossary		191
References		197
Index		205

Chapter 1

The History of Radio Astronomy

2009 was the International Year of Astronomy. This marked the 400th anniversary of the Italian astronomer Galileo turning his small refracting telescope towards the sky. Galileo observed the mountains on the Moon and the planet Jupiter with its four large satellites, which are now known as the Galilean satellites. He observed sunspots and made drawings of these sunspots and found that they moved across the Sun's surface. This led Galileo to note that the Sun rotates on its axis. Meanwhile, the history of unaided eye astronomy can be traced back thousands of years. All ancient civilizations looked up at the sky and saw patterns within the stars, and they gave these groups of stars names.

On this scale, radio astronomy is a relatively new science. It first had to wait until electricity had been discovered and understood, and until a way was found to control and use it. In the late 1880s a scientist by the name Heinrich Hertz conducted a series of experiments where he successfully transmitted and received radio waves. Hertz noted in his journal after he had demonstrated his ability to transmit and receive radio waves that "radio waves are of no practical use"!

1.1 The Early Years

The first attempt at receiving the newly discovered, radio waves from any part of the sky was in the early 1890s by Thomas Alva Edison. Edison thought it may be possible to detect radio waves from the Sun. He sent his laboratory assistant to the Lick Observatory in California with instructions on how to construct an antenna and receiver, from which he hoped to receive radio waves from the Sun. The receiving equipment and an antenna were built but no signals were ever reported as having been received. After this first failed attempt by Edison to receive radio waves from the Sun he made no further attempt in the future to repeat his experiment, which is covered in more detail in the chapter on the Radio Jove receiver.

Up to this point it was believed that radio waves could only travel in straight lines like the beams of light that come from a lighthouse. It wasn't until 1900 when Guglielmo Marconi demonstrated the first Trans-Atlantic radio transmission from Poldhu in Cornwall to Newfoundland that such long distance radio transmission was proved possible. It should be noted that there is some conjecture about whether Marconi actually did this or not, but that's for historians to prove or disprove. After Marconi made this transmission the idea that radio waves only traveled in straight lines had to be rethought and the idea of the mysterious medium which aided the propagation of electromagnetic waves, namely the "Luminiferous aether", were finally abandoned. Theories were then put forward that there must be some part of the upper atmosphere that was in some way able to reflect radio waves in order to allow them to travel the vast distances across the curved surface of the Earth.

In the meantime further attempts were made at receiving radio waves from the Sun. Between the years of 1890 and 1897 Sir Oliver J. Lodge built a more sensitive antenna and receiver in order to try and receive radio waves from the Sun, but unfortunately he also failed, whether it was due to his equipment not being sensitive enough or that there had been little or no activity on the Sun itself. The next attempt to receive radio emissions from the Sun was made by two astrophysicists, Julius Schiener and Johannes Wilsing. They constructed their own antenna and receiver, and this time they let their experiment run for just over a week, but still no radio emissions were received from the Sun. They incorrectly theorized that the atmosphere must therefore be absorbing all radio waves coming from the Sun. This added fuel to the theory of a reflective layer in the upper atmosphere that as well as reflecting radio waves back towards the Earth must also reflect any radio waves from space back into space. A further attempt was made at receiving radio waves from the Sun, but it was based on the incorrect theories developed from the week-long experiment by Schiener and Wilsing that the atmosphere must in some way absorb any radio waves trying to reach the Earth's surface.

Next, a French graduate student named Charles Norman took the interesting approach of taking his receiver and antenna up a mountain. At the altitude of over 3,000 meters (10,000 feet), he thought if Schiener and Wilsing were right about the atmosphere absorbing radio waves he should be able to at least receive something at this altitude, but unfortunately he also failed to receive anything. Experts looking back over his notes and his equipment have agreed that the equipment he had would have been sensitive enough to receive some radio emissions from the Sun but unfortunately had decided unknowingly to carry out the experiment at the time of solar minimum and it was just bad luck and not bad equipment that had caused him to fail. After all these attempts had ended in failure the idea of receiving radio waves from the Sun went out of fashion and it wasn't until World War One that this mystery would finally be solved.

Radio was still in its infancy and the equipment was large and bulky at the outset of the First World War, when the trench warfare system required a reliable and safe way for each trench to keep in contact with others. Radio equipment also required a fair amount of power to operate it, and any radio communication would have been easily intercepted by the enemy. The best and most reliable way to communicate was therefore to use a basic type of telephone and have wires running from trench to trench. The distance between each trench could be quite large, so long lengths of phone cables were needed. At certain times when the phones were being used strange hissing and whistling noises would be heard by the operators. This at first was thought to be an attempt by the enemy to intercept the communications, but this theory was soon dispelled, though a better understanding of the causes of these strange noises would have to wait until the end of the War in 1918. At that point it was realized that these noises could be made by very low frequency (VLF) radio emissions from the Earth itself. This will be covered in greater detail in a later chapter about the INSPIRE project.

Fast forward to the early to mid 1920s and extensive work had been done and was still being carried out by a number of scientists and researchers to prove the existence of this reflective layer high up in the Earth's atmosphere. Over several years radio signals were bounced off this reflective layer and the returning signals were studied in order to try and understand the layer's properties. This reflective part of the atmosphere, later to be known as the ionosphere, was found to reflect certain frequencies of radio waves like a mirror reflecting a beam of light. It was first thought that no radio waves could travel through the ionosphere. This was later proved incorrect and as the technology improved it was demonstrated that the ionosphere would indeed allow certain radio waves to pass through it, but they had to be of the right frequency.

A practice called "sounding the ionosphere" was introduced. This practice involved reflecting radar beams off the ionosphere from the Earth's surface to study its properties in order to understand its effects at different frequencies. Other attempts made to understand the ionosphere's properties included sending transmitters high into the Earth's atmosphere using high altitude balloons. Even rockets have been used, but this took place later, only after reliable rocket technology had been developed. This practice is still carried on up to the present day to study changes to the structure of the ionosphere and any changes due to seasonal variation. Today, satellites are also employed to help perform tests involving the ionosphere. In 1927 a scientist by the name of Sydney Chapman proposed a mathematical model of the Earth's ionosphere that was considered at the time to be the best model to explain the then known properties of the ionosphere.

Between 1930 and 1932 Karl Jansky, a telephone engineer working for Bell Laboratories in America, was tasked with looking into the problem of interference of long distance high frequency ship to shore communications at 20.5 megahertz. Bell Laboratories was looking into the possibility of making and receiving commercial Trans-Atlantic radio telephone calls instead of using telegraphs, where the wires for the telegraph equipment had to run along the bottom of the Atlantic ocean.

The problem was they kept encountering interference whose origin was not known, and could therefore not be counteracted. So, Jansky was given the task of identifying the cause of this unknown interference and to come up with a way to

Fig. 1.1 The merry-go-round antenna. Karl Jansky can be seen right of center (Image courtesy of NRAO/AUI)

eliminate it. Jansky built himself a large antenna which could be moved on a circular track. This contraption was affectionately nicknamed the "merry-go-round" antenna (see Fig. 1.1).

Jansky's antenna used two pairs of Ford Model T wheels to support the frame, allowing Jansky to move it single-handedly in the hope that through altering its direction he could narrow down the location of the interference problem. After running his equipment for a short time Jansky found that the interference seemed to come at regular intervals, every 23 hours 56 minutes and 4 seconds. This meant it happened nearly 4 minutes earlier each day. The timing must have left Jansky scratching his head as it didn't mean anything to him: Jansky was an engineer and not an astronomer. But as any astronomer would have known, the Earth doesn't rotate on its axis in exactly 24 hours, but takes 23 hours 56 minutes and 4 seconds, and thus a point in space such as a star rises approximately 4 minutes earlier each day. When Jansky realized this, he concluded that the source of the unknown interference must be coming from the sky and in fact from space itself.

Jansky tried to isolate which part of the sky the interference was coming from. This sounds easier than it actually is because as we will find out later, radio telescopes have very poor angular resolving power and it is very difficult to pinpoint an exact point on the sky using them. To identify the location of this radio emission

1.1 The Early Years

the best that Jansky could hope for was to track down the area of sky and possibly the constellation where the emission originated. He did this by estimating the width of the beam of his antenna and consulting star maps in order to know what part of the sky, or more importantly which constellation, was passing through the antennas beam at the exact time the interference was detected. He did this over a number of days, watching to see when the signal received by the antenna was at its strongest. This let him know that the origin of the radio emissions was at the center of the antenna beam. He quickly came to the conclusion that the source of the radio emissions or interference was coming from the part of the sky that contained the constellation of Sagittarius. The design of the antenna made it impossible to narrow it down any further than this. When Bell Laboratories received Jansky's report showing his results, they realized that the source of the interference was indeed of an extraterrestrial origin, and that although Jansky had done an excellent job, nothing could be done to solve the problem. Jansky was moved on to other projects.

Jansky quite rightly protested that this new extraterrestrial source should be investigated further as this was the first time such a signal of this nature had been received. He asked for extra funding in order to build himself a larger antenna and a more sensitive receiver to investigate this radio source further and try and determine the exact part of the sky where the radio waves were being emitted, but funding to build a larger antenna and more sensitive equipment was refused. Unfortunately, Karl Jansky died in 1950 from a heart condition at the young age of 44 after suffering with poor health for most of his life. In fact, it was due to this poor health that Jansky had been refused entry to the American Army. But in his honor the unit used in radio astronomy to represent the energy or "flux" coming from a radio source is called the Jansky. He also has a crater on the Moon named after him in recognition of the discovery of the first radio source beyond the Earth, though as this crater is on the far side of the Moon it is not visible from Earth. Up to the point that Jansky made his discovery it had not occurred to anyone that radio emissions could come from an extraterrestrial source.[1]

Karl Jansky published his findings in 1933 and this year would go down in history as the start of the science of radio astronomy. But nothing was done to follow this discovery to a conclusion until Grote Reber took up the challenge. As a young electrical engineer Reber was intrigued by Jansky's results about this mysterious source of extraterrestrial radio emissions, and was interested to follow up on Jansky's findings and take them to what Reber describes as a "logical conclusion". Reber applied several times to Bell Laboratories for a job with the intension of following up on Jansky's pioneering work, but his applications were refused not because of his qualifications or his intention to follow up on Jansky's work, but this was the mid 1930s and America was in the middle of the Great Depression and jobs

---

[1]There is an interesting podcast about Karl Jansky at www.365daysofastronomy.org (dated 11 April 2012). This podcast was written and recorded by Dr Christopher Crockett of the United States Naval Observatory, as part of Dr Crockett's astronomy word of the week series. The podcast is well thought out and presented and worth tracking down and listening to.

Fig. 1.2 Grote Reber's homemade radio telescope (Image courtesy of NRAO/AUI)

were few and far between. Reber, being of strong character and not wanting to be beaten, decided to build his own radio telescope (see Fig. 1.2).

The antenna, as can be seen, is of a parabolic dish design and was in fact the first antenna purposely built to study the sky and therefore the very first radio telescope. The diameter of the dish is 9 meters (29.5 feet). Reber built this by himself in his own garden, constructing the whole telescope frame and shaping the dish to give it a parabolic shape. He also built all his own receivers. This was long before the miniaturization of electronics with the use of transistors and microprocessors. The receivers would have been made up of a collection of vacuum tubes (valves) used in the amplifiers and receiver circuits. Vacuum tubes are very fragile, being made of glass, and difficult to manufacture, so they would have been quite expensive. Vacuum tubes also demand more electrical power to operate and take time to "warm up" before they will operate properly.

Reber knew that Jansky received his signals at 20.5 megahertz, but Reber would have had to operate his radio telescope at the frequency at which the parts that were available to him dictated.

Working alone, Reber started building in 1933 and when he had finished he had a radio telescope that could only be moved in altitude (up and down) not in azimuth (east or west). This meant that the rotation of the Earth was responsible for the east to west movement of the telescope. After he had built his radio telescope Reber started to make sweeps of the sky, but he found that the interference from the local neighborhood, especially from the sparks of vehicle ignition systems, made this very difficult. He tried all different times of the day and found that the best time to observe was from midnight to 6 am once everyone was in bed asleep. Between these times things quieted down enough to allow him to make useful observations with his equipment. This optimum time frame for radio astronomy is still true today and meant that Reber's day of full-time work at a radio factory was followed by his evening meal and then a short nap, until just before midnight he would rise and start observing with his radio telescope from around midnight to 6 am. Then, once again, he would eat breakfast and drive to work. And so it went on night after night. He did this for a number of years until he had built up the first radio map of all the sky that he could observe from his location. Now that's dedication!

Although his map would be considered rather crude by today's standards, Reber did confirm Jansky's findings of a radio source in the constellation of Sagittarius. He also found strong radio emissions coming from other areas of the sky and published his findings in 1938. Later, the American National Radio Astronomy Observatory (NRAO) research center based in West Virginia employed Reber as a consultant from the early 1950s. Reber donated his homemade radio telescope to the NRAO and it is now sited at the NRAO's Greenbank Science Center where it is preserved and maintained. It also has been fitted on to a turntable which allows it to be turned in azimuth, so it now has a true altazimuth mount. In 1954 Reber moved to Tasmania and worked at the University of Tasmania, where he carried on with the study of radio astronomy. He died a little before his 91st birthday. Those who knew him said that his mind was as sharp as ever up to his death but it was his body that had given up on him.[2]

1.2 Wartime Developments

At the outbreak of World War Two in Europe in 1939 a veil of secrecy descended over most of the world that was affected by the War. All communications and messages would be monitored between the countries involved in the fighting. Also, the countries taking part could ill afford luxuries like funding research into things like

---

[2]There is a very good podcast available on the internet which includes an interview with Grote Reber himself. The podcast can be found at the Mountain Radio website http:/www.gb.nrao.edu/epo/podcasts.shtml. Reber is interviewed by a member of the Greenbank Radio Telescope staff. Reber talks about how he built his radio telescope and the problems that he had to overcome in doing so, and how everyone thought he was an eccentric and a bit of a crank. He tells the story of how he would be supplied with boxes of vacuum tubes from the manufacturer and would spend hours going through them all in order to find the correct piece to do the job he wanted it to do. It's a shame this podcast wasn't longer as he makes for pleasant and easy listening.

radio astronomy. All the scientists in these different countries found themselves in the middle of things, with a great deal of academic research being put on hold as they were given new duties such as code breaking. The only research that was allowed was trying to come up with weapons that would give their country the edge over their enemy. The United Kingdom had developed a new invention which was later to be called RADAR, for "Radio Direction And Ranging". The first attempt at doing this was very basic. Tall wooden masts with wires strung between them transmitted a radio signal and a separate mast and antenna were used to receive the returning echo. The returning echo was shown on an oscilloscope screen and the operator had to make their best guess regarding what the signal meant and how far away an object was.

As WWII raged on the call for better radar equipment increased and a number of radar installations were built to give radar cover to the entire east coast of the United Kingdom. All the radar installations were in contact with each other by telephone. The advantage of having a larger number of these installations was that each station could compare their signals to the ones on either side to see who had the strongest echo. This information could then be used to give a better and more accurate way of judging an object's distance and direction of travel. It was the job of the station with the strongest echo to call the results in to the senior person so a decision could be made on what would be the best course of action to take. Although this system was very basic it served to give enough warning so that defensive action could be taken while the enemy planes were over the sea, where if shot down the loss of civilian life was limited.

Some British radar operators reported receiving a large echo within their equipment that seemed to be coming from over mainland Europe. In 1942 a group of technicians led by J. Stanley Hey was sent to find the origin of this signal, just in case it was an attempt by the enemy to block or jam UK radar. After an exhaustive search and a thorough examination of the equipment, Hey and his colleagues thought that the Sun could be the source of the signal that the radar operators had reported receiving. Astronomers observing the Sun confirmed that a large group of new sunspots had appeared on the Sun's disc. This then made the connection between solar activity and radio emissions from the Sun. Hey didn't publish his findings until 1946, after the end of the War, but during the War Grote Reber had also detected emissions from the Sun and published his results in 1944. This will be covered in greater detail in the chapter on the Radio Jove receiver.

Also invented by scientists during the War a new piece of equipment called the cavity magnetron. This was a key development which gave rise to a more accurate radar unit using shorter centimeter wavelengths. By using these smaller wavelengths smaller objects could be found and tracked more effectively, as the smaller wavelengths would readily reflect off an object. This meant that radar units could also be made smaller and even that they could be made small enough to fit into an aeroplane, giving rise to airborne radar.

This progress came at a price. In the rush to develop the cavity magnetron safety was not as high a priority as maybe it should have been, and quite a few people were irradiated, some quite badly. In the workshops where the cavity magnetron

1.2 Wartime Developments

was developed people reported that the cheese in their sandwiches started melting, even at some distance from the radar unit, every time it was switched on. When the radar units were finally fitted into aeroplanes there was very little in the form of screening from these new radar units and some of the flight crew reported that bars of chocolate they had been carrying with them had also melted quite badly, providing further evidence that the screening was inadequate. The workings of the modern microwave oven are in fact a spin-off of the technology used to develop the cavity magnetron, but the screening problem surrounding the leaking of microwave energy has now been adequately addressed (well, one hopes it has).

New radar equipment developed in WWII could be made smaller and more efficient. Portable anti-aircraft units were now available, mounted on a trailer unit that could be towed to almost anywhere they were needed. While the radar operators were using this new portable equipment they reported receiving short duration echoes which lasted from less than a second up to around 5 seconds. This was another mystery that had to be solved. By this time in the war, German V1 and V2 rockets were being sent over the Channel to England from mainland Europe. These V2 rockets travelled at high velocities approaching supersonic speeds. In fact, the technology from these V2 rockets would eventually put astronauts on the Moon. Had these short duration echoes anything to do with these new weapons? J.S. Hey and his group came to the rescue again. Hey and his colleagues looked into all possibilities that might explain what the mysterious short duration echoes were. They came up with the theory that these echoes could be coming from meteors as they enter the Earth's atmosphere.

After the end of World War Two there were lots of spare ex-army portable radar units being sold off cheaply in order to try and recoup some of the money that the War had cost to fight. Hey got hold of some of these units and carried out experiments using the radar units to prove his theory that these short duration echoes really did come from meteors entering the Earth's atmosphere. In 1945 he did indeed prove that these short duration echoes originated from meteors entering the Earth's atmosphere. This will be covered later in the chapter on meteor detection.

Hey was just one of a number of scientists at the time who wanted to use radar equipment for a more useful and productive purpose. Across the pond, the birth of the American space program started with Project Diana in the mid 1940s. This was to be the first US government attempt at radar astronomy. At the time the ionosphere was still a relatively new concept and its properties weren't yet fully understood. No one really knew if it was possible for a radio signal to penetrate the ionosphere, and if this was possible at what frequency. Project Diana was an attempt by the United States to try and bounce a radar signal off the Moon using a modified Second World War radar transmitter and receiver; the name Diana was chosen for the project because the goddess Diana in Roman mythology was said to hunt animals at night only by the light from the Moon.

A site in New Jersey containing a transmitter and receiver was constructed for the project. The radar equipment used was a SCR-271 early warning radar unit and was army surplus left over after World War Two. It had been modified by Major Edward H. Armstrong. Armstrong was an army consultant during the War and was famous for pioneering the use of frequency modulation. The transmitter was designed to transmit at a frequency of 111.5 megahertz in 0.25 second pulses. It was hoped that the power of the transmitter, some 3,000 watts, would be enough to send the pulses through the ionosphere and all the way to the Moon and hopefully back to the Earth. The antenna had limited movement and could only be moved in azimuth, which meant scientists had to wait for the Moon to be in the right part of the sky and within the beam of the transmitter before any attempt could be made. The first successful detection of an echo was by John H. Dewitt, Jr. and E. King Stodola in January 1946. The received echoes returned approximately 2.5 seconds after transmission which travelling at the speed of light would be about right for a trip to the Moon and back.

Meanwhile in 1945, Bernard Lovell had returned to his post at Manchester University in the UK after his own wartime service working on the development of radar. He wanted to carry on with research into cosmic rays, highly energetic particles that enter the Earth's atmosphere and are now thought to originate from some of the most violent explosions in the universe, such as those of supernovae and hypernovae (super-supernovae). Lovell had the theory that meteor echoes that were received by the portable military radar units may in fact be cosmic rays entering the Earth's atmosphere, and the radar may be a way in which to study them.

Radar observations and any other type of radio observations taken from the city of Manchester were all but useless because of the level of interference from the inhabitants of the city, so a more remote and radio quiet observing site needed to be found for Lovell to make his observations. Luckily for Lovell, Manchester University rented a plot of land approximately 32 kilometers (20 miles) to the south of the city. This area, named Jodrell Bank, was used by the University as a botanical garden. A team was duly sent to Jodrell Bank with a portable radar unit in tow. At that time the site was nothing more than a large open expanse of land. The radar unit was fitted with a Yagi antenna which was mounted onto the receivers' trailer. This was then set up at a convenient area at the side of a wooden building that at the time housed the agricultural workers' tools. This wooden building would become the headquarters for the equipment and staff whilst they were at the Jodrell Bank site.

By the December of 1945 the team at Jodrell Bank had confirmed that Hey was right and that the radar echoes did in fact come from ionized meteor trails, and not as Lovell had first thought from cosmic rays. Lovell realized that he would need a more sensitive radio telescope if he was going to detect cosmic rays. In 1947 the researchers at the newly named "Jodrell Bank Experimental Station" built a 66 meter (218 foot) parabolic reflector, made of wire mesh, which pointed up towards the sky. There was a mast in the center of the wire mesh reflector that could be moved around by the tightening or slackening of a number of guide ropes used to support this central mast. This mobile central mast meant that the radio telescope was slightly steerable a few degrees or so in either direction from the zenith. It must have been a nightmare to try and point this central mast in any particular direction with any accuracy by the tightening and slackening of a few guide ropes, but over a period of time the team built up, strip by strip, a map of the sky over their heads as the Earth rotated.

Lovell never did detect any echoes from cosmic rays, but the telescope made the first detection of radio emissions from the Andromeda galaxy. This proved the existence of radio emissions outside our own galaxy of the Milky Way. Lovell, pleased by the results from the telescope but frustrated by the limited steering and the inability of the telescope to cover a greater area of the sky, put forward plans to design and build a fully steerable telescope with a larger collecting area or dish. The new radio telescope design was to have a telescope with a wire mesh parabolic dish with a diameter of 76 meters (250 feet). The whole structure was to be mounted on its own track so that the entire assemblage could be turned in azimuth. Either side of the receiving dish was to be a tower, and each of these towers would have an electric motor and a huge bearing inside a room at the top. These bearings would support the entire weight of the receiving dish assembly and would allow the dish to be pointed at any angle in altitude, thus giving the radio telescope a huge altazimuth mount.

1.3 Increasing Resolution

Radio telescopes have one large drawback compared to optical telescopes, and that is the problem of very poor resolution. This can be in the region of 100,000 times worse. This poor resolution is down to the fact that radio telescopes work at longer wavelengths, somewhere in the region of 100,000 times longer than those used by optical telescopes. To put this in perspective, to get the same resolution with a radio telescope as an optical telescope with a 2 meter (78.7 inch) mirror the radio telescope would require a parabolic dish measuring 200 kilometers (124.3 miles) in diameter.

The construction of a radio telescope is also very different from that of an optical telescope. The mirror of a large telescope has to be made to very high specifications and must have the correct shape in order to bring all the wavelengths of visible light to the same focusing point. For example, the mirror of the Hubble space telescope was slightly misshapen by a tiny fleck of paint that got in somewhere it shouldn't have during the grinding of the mirror. This tiny imperfection in the mirror was still enough to make the telescope "short sighted". NASA only managed to repair this fault by fitting a number of corrective lenses between the light coming from the mirror and the cameras and other instruments. Simply put, the astronauts fitted the telescope with eyeglasses.

On the other hand, for a radio telescope's collecting area the equivalent of a mirror can be made of solid sheet metal bent to form a parabolic dish, and in some cases this metal can be in the form of a wire mesh. No one in their right mind would try and make an optical telescope's mirror out of wire mesh, as visible light is of a small enough wavelength to pass straight through the mesh, but because of the greater wavelengths used in the radio part of the spectrum an incoming radio wave doesn't 'see' the holes in the mesh and will just be reflected. The shape of the collecting area is still important to bring the radio waves to a common focus, but this is why radio telescopes have such poor resolution compared with their optical counterparts.

One way around the problem of poor resolution in radio telescopes is to use them to form what is known as an interferometer, first used experimentally by Martin Ryle at Cambridge University in 1946. Interferometry is the practice of using two or more smaller radio telescopes to mimic a single much larger radio telescope. As with optical telescopes, the larger the collecting area the better the resolution. Unlike optical telescopes that work at a higher wavelength where it is much easier to pinpoint an object's position within the sky, a radio telescope's resolving power is quite poor in comparison, simply because of the longer wavelengths that are used. If two radio telescopes are used at a distance apart this gives the same effect as having one larger telescope. By increasing the distance between the two smaller radio telescopes, known as the "baseline", the resolving power of the two smaller telescopes can be greatly increased. As this baseline is increased, the gain in resolution of the two or more antennas is also increased.

For example, it would be a technological nightmare to try and build a radio telescope with a parabolic dish measuring, for instance, 200 kilometers (124.3 miles) in diameter. But it is quite feasible to build a number of smaller dishes that could be placed at intervals that add up to a collecting area of the 200 kilometers. Then, if the signals from each individual telescope dish are combined, this would give the same gain as the one huge dish.

The quality of resolution for any particular interferometer system has a number of factors that need to be taken into account. The first is the distance at which the telescopes are placed from each other (the baseline). The second is the frequency at which the system is operating. If we now look at the very long baseline array (VLBA) which is made up of ten radio telescopes that have dishes measuring 25 meters (82 feet) in diameter and that are spread across the continental United States and Hawaii, this gives a baseline of many hundreds of kilometers, with a combined resolving power approaching 0.001 arc seconds. This resolution of 0.001 arc seconds for the VLBA now outperforms the resolution of the largest ground-based optical telescopes. This is due to the effect of the atmosphere that limit optical telescopes to about 0.3 arc seconds. With the modern adaptive optics systems that are now being fitted to new ground-based optical telescopes it is even possible for a ground-based optical telescope to match the resolving power of the Hubble space telescope by flexing the mirrors of the telescope to cancel out the effects of the Earth's atmosphere.

Interferometry's technical clarity is the way forward, but as always there is a down side: if all these individual dishes cannot be coupled together for whatever reason, then when the received signal is recorded this recording must be extremely well time coded. Time codes act like indexing marks. Think of them as "pips" on the speaking clock, and when each recording is played back the time codes on the recording allow the different recordings to match up with the other telescope recordings, adding them together to give an improved signal. Timing these recordings to the nearest second is by no means good enough. The timing must be accurate at least to the nearest millisecond or better. Jodrell Bank in the United Kingdom uses an atomic clock which is accurate to one second every 25 million years to time code their observations. This system allows weaker signals to be identified against the background noise from space.

1.3 Increasing Resolution

From the mid 1940s to the late 1950s several surveys of the sky were carried out using a number of different frequencies by a number of radio telescopes around the world, identifying several areas of the sky from which radio emissions had been received for further study. These new areas of interest were called "radio stars". The term "radio star" is very rarely used in astronomy today and has been largely forgotten. But in old astronomy books the term crops up every now and then. Radio star was the name given to the area of the sky from which radio emissions were received. The key word here is "area", as the larger wavelengths used in radio astronomy made it very difficult to pinpoint the exact point in space the radio emissions coming from. With radio astronomy being a new science at the time no one had any real idea what was producing these radio emissions.

Until the accidental discovery of the galactic center by Jansky in 1932 some thought that radio waves could only be produced by manufactured transmitters. The hunt was on to find an accurate explanation of what was producing these radio emissions. Many maps were drawn showing the relative position of these mysterious radio stars. Drawings were also made to show the distribution of these radio stars within the galaxy in a hope this would in some way help towards an explanation, but no one really knew what they actually were and if they were all connected with our galaxy or not. This situation continued until Martin Ryle and F.G. Smith carried out research into Cassiopeia A, Taurus A and Cygnus A at Cambridge in the 1940s. These radio sources had already been reported by Hey and Bolton, and Ryle and Smith were given the task of finding an accurate position for the three signals. For equipment, the pair formed an interferometer using two 8.2 meter (27 foot) parabolic dishes from two old German radar units paired together.

As explained earlier, interferometers give greater accuracy for estimating an object's position and provides hints as to its overall angular size. Ryle and Smith found the source of Taurus A to be the area around the Crab nebula. This was known to be a supernova remnant, but no reason was found as yet regarding where the radio emissions originated from and what was producing them. When an accurate position was found for Cassiopeia A optical astronomers were given the coordinates and asked to observe this area to see if anything could be seen in optical light. Optical astronomers reported seeing a faint nebula. As a result of this faint nebula being seen, astronomers Walter Baade and Rudolph Minkowski at the Mount Palomar Observatory in California were asked to observe the area and double check the coordinates using the observatory's 5 meter (200 inch) optical telescope. A photographic plate was taken of the area using the same coordinates, and the existence of the faint nebula was proven to be correct. The two astronomers at Mount Palomar managed to extract a spectrum of the faint nebula and examinations of the spectrum found that it contained some quite unusual characteristics, making Cassiopeia A the first radio source of its type to be confirmed by optical observations. After the faint nebula's redshift had been calculated from its spectrum it was thought to be located at a distance of 10,000 light years. This distance places the nebula still within the Milky Way galaxy.

Grote Reber was the one to initially find the radio source Cygnus A, appearing on the radio map of the sky in 1944. Hey knew this area had a nebula that measured several degrees across (NGC 6992 the Veil Nebula) thought to be the result of a supernova. Cygnus A took quite an effort to track down with any accuracy because the nebula that is the result of the supernova covers quite a large area, and because the area is dense with stars as the Milky Way passes through this part of the sky. When Cygnus A was finally tracked down and photographed the resulting image showed the nuclei of two colliding galaxies and proved that the radio emissions were coming from beyond the visible Veil nebula. When a spectra was taken and the redshift calculated the two colliding galaxies were thought to be at a distance of 550 million light years, making Cygnus A up to this time the furthest extragalactic radio source identified outside our own galaxy.

In 1951 Harold I. Ewen and Edward Mills Purcell discovered the 21 centimeters line of hydrogen, allowing another leap in radio astronomy. The scientists used a horn-shaped antenna of approximately 2 meters (72 inches), square at the open end and reducing in size along its length like a large square ice-cream cone. The 21 centimeters hydrogen line is a change of energy state of neutral hydrogen atoms that happens at a precise frequency of 1,420.40575177 megahertz, equivalent to a wavelength of a little over 21.11 centimeters. The physical change that happens to the hydrogen atom is that there is a change in "spin" of the orbiting electron. This "spin" quality is the property of electrons and other fundamental particles and how they rotate about their axis. Simply put, a hydrogen atom has one proton and one electron, and the "spin" of the proton and electron can either be the same or oppose each other. If the "spin" of the electron opposes the "spin" of the proton the total energy within the atom is slightly less than it would be if the two had the same "spin". At the moment the electron changes its "spin" the energy lost from the atom has to go somewhere, as per the law of energy conservation, so a low energy photon is released. The energy level of this photon is equivalent to the wavelength of 21.11 centimeters.

This discovery held the possibility of studying the spiral structure within the Milky Way. As the hydrogen wavelength can pass through clouds of interstellar dust that the visible wavelengths cannot, astronomers could now observe what was happening inside gas and dust clouds. The discovery of the 21 centimeters hydrogen line had a profound impact on the scientific community and also forced a change in the design of the radio telescope proposed at Jodrell Bank. The change in design involved replacing the original wire mesh dish with a solid dish. This new solid dish with its smoother surface would allow the telescope to be more efficient, and would also allow it to work at a greater range of frequencies including the 21 centimeters hydrogen line. But before construction of this new radio telescope at Jodrell Bank, named the Mark 1 radio telescope, could start, there were problems over where to site the radio telescope due to the sheer size and weight of the finished structure. It was estimated that the finished telescope would weigh in the region of 1,700 tons.

There were also the inevitable financial problems to contend with now that the design had been changed from a wire mesh to a solid metal dish, as having a solid metal dish added considerably to the total cost. The project very nearly came to a

1.3 Increasing Resolution

Fig. 1.3 The Lovell radio telescope (formally the Mark 1) at Jodrell Bank Cheshire in the United Kingdom (This image was taken shortly after the surface of the dish had been recently replaced with a more accurately shaped and smoother surface which allowed the telescope to be used at even higher frequencies. Also at the time the image was taken the structure was having a fresh coat of paint)

standstill on several occasions, and a plea for money to finish the Mark 1 radio telescope was issued to anyone and everyone. There is a charming story of a young school child who wrote a letter to Bernard Lovell upon hearing about the plight of the unfinished telescope. In that letter the school child expressed their concern over the fate of the telescope and the child's belief that the telescope should be built in order to explore the universe, and included spending money within the letter as a contribution in the hope that the telescope would soon be finished. This would have surely tugged at the heart strings of Bernard Lovell, and perhaps it made him redouble his efforts to complete the telescope. By 1957 the Mark 1 radio telescope was finished, the world's largest fully steerable radio telescope at the time. Thirty years later the telescope was renamed the Lovell telescope in honor of Sir Bernard Lovell. References will be made to the Mark 1 radio telescope several times later in the book as it crops up often within the history of radio astronomy (Fig. 1.3).

By the middle of the 1950s radio astronomy had really taken off as scientists all around the world had started building all manner of weird and wonderful radio telescopes to explore the universe in radio wavelengths. Between the 1950s and the 1960s there were several large radio telescopes built at Cambridge and Jodrell Bank in the United Kingdom, Westerbork in The Netherlands, Parkes in Australia, Greenbank in the United States, and Arecibo in Puerto Rico to name a few. It was in this environment that the next discovery in radio astronomy that took scientists completely by surprise occurred, in 1955.

1.4 Planetary Signals

Astronomers Bernard F. Burke and Kenneth L. Franklin of the Carnegie Institute were testing out a new antenna and working at a frequency of 22 megahertz when they accidentally discovered radio emissions coming from the planet Jupiter. After the antenna was used for the first time, Burke and Franklin had the usual problem of working out the signals they wanted from the unwanted interference. They worked their way through each of the wanted and unwanted signals, but there still remained a frustrating bit of "scruff" that couldn't be identified. After an excellent piece of detective work it was found that this bit of scruff that was thought to be coming from vehicle ignition systems was in fact the planet Jupiter. This was the first planet ever detected at radio wavelength.

Radio astronomy has been useful in observing other planets as well. Venus, named after the Roman goddess of beauty, is the third brightest object in the sky after the Sun and Moon but has always been a bit of a mystery. The planet is shrouded in a thick blanket of cloud and no views of the surface can be seen from Earth-based optical telescopes only details in the cloud tops can be seen. Using ultraviolet wavelengths can help in seeing some detail within the Venusian cloud tops, but this wavelength has its own problems as some ultraviolet wavelengths are blocked by the Earth's atmosphere. Because Venus is closer to the Sun than the Earth is, it seemed reasonable to assume that the average temperature on Venus would be higher than those of the Earth. Many estimates were put forward regarding what this temperature at Venus's surface could be. In 1952 Harold Urey made a prediction that the surface temperature of Venus would be around 53 degrees Celsius (127.4 degrees Fahrenheit). Spectroscope analysis of the atmosphere showed that there was a high level of carbon dioxide present, but at this time the effects of such a high level of carbon dioxide in the atmosphere were not fully understood, and it was long before the phrase "greenhouse effect" had even been thought of.

One British astronomer, on hearing about the high levels of carbon dioxide, suggested that the oceans of Venus could be fizzy like carbonated water with all the dissolved carbon dioxide in them. In 1956 a group of radio astronomers working at the Naval Research Laboratory in the United States, one of whom was C. H. Mayer, studying microwave radar emissions received back from Venus and suggested the temperature could be closer to 330 degrees Celsius (626 degrees Fahrenheit). The Russians would find that this was closer to the actual temperature of around 470 degrees Celsius (878 degrees Fahrenheit) when they started sending the series of Venera space probes to the planet in the early 1960s, ignorant at that time about the crushing atmospheric pressure on Venus, 90 times greater than the atmospheric pressure of the Earth. This would prove to be a serious problem as several of the early Venera probes were crushed by this pressure even before reaching the surface of Venus.

1.5 Sputnik

Just months after the Mark 1 radio telescope at Jodrell Bank was finished the Russians planned to launch a series of satellites into Earth's orbit. These were to be the first artificial satellites to orbit the Earth. These were the Sputnik series of satellites, and it was a good test for the newly finished Mark 1 radio telescope at Jodrell Bank to monitor these satellites in orbit to test the telescope's accuracy and steering capability. On October the 4th 1957 the launching by the Russians of "Sputnik 1", the world's first artificial satellite, started the space race, and the finishing line of this race was going to end at our natural satellite the Moon (Sputnik means "fellow traveler" in Russian). The satellite Sputnik 1 was spherical in shape with a diameter of approximately 580 millimeters (23 inches). It weighed approximately 84 kilograms (185 pounds). As it flew over the Earth its distances varied from 240 kilometers (150 mile) to 480 kilometers (300 miles) according to its elliptical orbit and on board was a simple transmitter that allowed scientists to track Sputnik's orbit around the Earth. The satellite transmitted on two frequencies, 20 and 40 megahertz. Some think these particular frequencies were deliberately chosen in order to serve two purposes. The first was to study which frequency passed through the ionosphere best – whether the higher 40 megahertz or the lower 20 megahertz. It can only be assumed that this was a test to choose a suitable frequency for a manned mission later on. The second reason was at the time of the Sputnik 1 launch, it didn't take much imagination to retune a household "wireless" and tune into its "beep beep" signal as it orbited overhead, thus proving to the world that the satellite was really in orbit.

On November 3rd of that same year, Sputnik 2 was launched and the knowledge gained by the Sputnik 1 test regarding what frequency to use to transmit through the ionosphere was put to use. This time a small 3 year old female cross breed dog called "Laika" was on board. Found roaming the streets as a stray, Laika was selected from two other dogs and had to endure a punishing training schedule. Before launch Laika was fitted with a harness that restricted her movements within the capsule, she also had sensors fitted to monitor her heart rate and other vital signs. On entering orbit the radio receiver on the Earth was switched on and barks could be heard coming from the capsule. Laika had survived the journey into orbit, and at this point everything seemed to be going well. But this was a short-lived victory as she died shortly afterwards, within a matter of hours. There are many theories why she died, but it is now generally agreed that she died from heat exhaustion. As the temperature within the capsule started to rise, possibly due to damaged insulation on the capsule or a failure in her life support system, no one will ever know for sure, her vital signs started to show that she was becoming more and more distressed. Then suddenly they started to drop until there was no pulse rate being received back on Earth. Her death caused outrage and a great deal of controversy throughout the world, but the Russians responded by saying it was always intended for her to die as no plans were made to bring her back to Earth. This response just made things worse, and the whole incident overshadowed the Russian's part in the space race to the Moon. Laika will always be known as the first living creature in space, and also the first to die in space. In April of 1958 the capsule that contained her remains burned up as it re-entered the Earth's atmosphere.

Now the Russians had proven that it was possible for a living creature to survive in space, albeit for only a few hours in the case of Laika, and that communication was possible between a spacecraft and the Earth through the ionosphere. The space race was truly on, and it was only a matter of time before the first human being would orbit the Earth and then travel to the Moon. Both the Americans and Russians started sending unmanned robotic probes to the Moon in order to get close-up images, with the aim of looking for a suitable landing site for a manned mission. The Americans had the Ranger series of probes and the Russians had the Lunik series of probes. Both countries were hoping for close-up images of the Moon as the spacecraft crashed headlong into the surface of the Moon. The aim of this was to study the composition of the Moon's surface, and address vital questions such as whether the Moon would support the mass of future astronauts or if they might just sink into the surface never to be seen again.

In 1959 the Russians launched the Lunik 3 probe. Its mission was to orbit the Moon and take images of the far side of the Moon. This was of interest as no one had ever seen the far side before, the Moon's orbit being tidally locked to the Earth. At this time the Russians didn't have an antenna capable of communicating with the Lunik 3 probe at the distances that would be involved as the probe travelled to the Moon. So Jodrell Bank was asked to use the newly built Mark 1 radio telescope to do the job for them, which they did. As the Lunik 3 probe went round the back of the Moon, radio contact was lost and all that could be done was wait and hope until it reappeared from behind the Moon. As Lunik 3 came back into contact after its trip around the far side of the Moon the Mark 1 radio telescope started to receive a signal. Bernard Lovell recognized the incoming signal as a facsimile data signal, at the time this was the only technology available for sending an image via radio communication.

The only establishments at the time to have a fax machine that was capable of turning this incoming signal back into an image were the fax machines used by newspaper offices. A telephone call was made to a Manchester newspaper office and a fax machine was quickly brought to Jodrell Bank, where it was connected to the radio telescope's feed. As the image was built up line by line it wasn't long before they realized that it was an image from the far side of the Moon. The following morning the image of the far side of the Moon was printed on the front of the newspapers for all the world to see. The Russians were quite understandably not very happy about the fact the world had seen the images before them, and the photograph very nearly caused an international incident between the United Kingdom and Russia.

1.6 The Search for Intelligent Life

In 1959–1960 a radio astronomer named Frank Drake suggested that if all radio telescopes around the world were listening to the apparent random static from space that it would be a good idea to listen just in case an extraterrestrial civilization in a far off star system was trying to contact the Earth using radio waves, broadcasting to any other civilization who was in the path of their signal and had the technology to receive and understand the incoming signal. Drake theorized that this extraterrestrial signal could be either a deliberate attempt at contact across the vastness of space, or an accident caused by a radio telescope picking up "leaked" electromagnetic radiation from an extraterrestrial civilization's communications in the same way that Earth's radio and television signals have been leaching into space since the first radio and television transmissions were made.

This idea led to the birth of SETI, "The Search for Extraterrestrial Intelligence". Frank Drake carried out the first of these SETI searches in 1960 at the time he was at the National Radio Astronomy Observatory (NRAO) in Greenbank, West Virginia in the United States. This project was called project Ozma after the queen of the fictitious land of Oz from the movie "The Wizard of Oz". To start the program, Drake had to first resolve the problem of whereabouts in the sky to point the telescope to give him the best chances of receiving a signal. He decided to aim towards a star that was roughly the same type and age as our own, his reasoning being that our Sun is a long lived and stable star that has existed long enough for advanced life to have developed on a planet orbiting it, and if planets did exist around other stars then a star like our own would be a good place to start.

The next problem was to find suitable stars at the right distance. Radio waves have been leaking into space from the Earth for just over 100 years. As radio waves travel at the speed of light the greatest distance they could have travelled is just over 100 light years. So, if an extraterrestrial civilization was trying to do the same experiment as Drake and listen in to the radio emissions from the Earth, and if the extraterrestrials were 150 light years away, then they wouldn't have the faintest idea that the Earth was emitting signals for another 50 years. Thus Drake picked stars that were close to the Earth in order to give him greater odds at receiving a signal. The third and final problem was to determine what frequency he was going to tune his receiver to. This could be almost anything, and if Drake chose the wrong frequency then the extraterrestrial civilization could be beaming out radio signals and Drake would have no idea that they were there. Drake asked himself if he were going to transmit a radio signal what frequency would he choose and why. He theorized that the frequency had to be something that was the same throughout the universe and would be well known to an advanced extraterrestrial civilization, and so he selected a frequency that was associated with the most abundant element in the universe, that of hydrogen and the 21 centimeters line at the frequency of 1,420 megahertz.

Using those parameters, Drake selected the stars Tau Ceti in the constellation of Cetus the Whale and Epsilon Eridani in the constellation of Eridanus the River.

Both stars were of an age similar to that of the Sun and both were around 11 light years away in distance. Drake had chosen which stars he was going to observe at what frequency, now all he had to do was listen and wait. From the months of April to July of 1960 he tuned the 26 meter (85 feet) NRAO telescope to 1,420 megahertz and for a period of 6 hours a day he recorded the output from the receiver and waited for some sort of signal.

Drake and the other astronomers involved repeatedly played the recordings they had made, looking for any signal that would lead them to believe they had received an extraterrestrial signal, but nothing was forthcoming. They tried looking for codes within the recordings with mathematical connotations such as prime numbers and patterns within any pulses that seemed to repeat themselves at regular intervals, but still they found no evidence of a signal. After Project Ozma had taken the first brave steps in looking for extraterrestrial radio signals from alien worlds it made it a feasible scientific goal. Later, Drake came up with his famous equation:

$$N = R * Fp\ Ne\ Fl\ Fi\ Fc\ L.$$

N = the number of civilizations within our galaxy that communication with, maybe possible.
$R*$ = the average rate of star formation in our galaxy per year.
Fp = the number of stars that may have planets orbiting them.
Ne = the number of planets around a star that could support life.
Fl = the number of planets above that go on to develop life.
Fi = the number of planets that develop intelligent life.
Fc = the number of civilizations that develop the technology to make and receive a signal or sign to acknowledge their existence.
L = the length of time that such a civilization produces their signal or sign.

This equation was a list of variables that were thought to be important to work out the probability of there being a civilization with the technology to make and receive a radio signal or other sign to make their existence known. There is no right or wrong answer to the equation, as the variables can only be estimated with the best knowledge at the time. What was needed was a systematic search of all the stars that met all the requirements that Drake had first thought of, including long-lived, stable stars that have existed long enough for planets to have formed around them.

1.7 The Apollo Mission

Drake's research occurred at a time when radio astronomy was looking not only at distant stellar systems. As the 1960s dawned, the President of the United States of America John F. Kennedy made his famous speech about sending a man to the Moon before the decade was out, and returning him safely back to the Earth.

This presented NASA with a huge task. Not only had they got to come up with a way to get the astronauts to the Moon and back again, they would also have to find a way to communicate with them around the clock while they were travelling to and from the Moon, and whilst they were on the surface of the Moon. This was a problem because no single antenna, no matter how big, could communicate with a spacecraft going to the Moon if the Earth was facing away from the Moon when a transmission needed to be made or received.

What was needed was a number of antennas situated around the world and spaced at suitable intervals. The idea was that as the Earth rotated and each antenna lost sight of the Moon as it set below the horizon at that location another antenna could take over the communication with the spacecraft, because the Moon would be rising at that location. If done correctly, and if a small overlap between antennas were possible, this overlap need only be a few minutes to allow one antenna to lock on to the signal from the spacecraft before the other antenna lost it, then the changeover could take place seamlessly. This required the cooperation of a number of countries, and several suitable antennas around the world were found or built in what would come to be known as the NASA Deep Space Network. By the start of the Apollo missions NASA had its Deep Space Network up and running and round the clock communications with spacecraft travelling to and from the Moon were now possible. There is a movie available based on these true events called "The Dish" (Houston's other problem) that tells the story very well.

The three Apollo 11 astronauts, Neil Armstrong, Buzz Aldrin and Michel Collins, approached the Moon and planned to go into orbit around it. Everything with the NASA Deep Space Network was working fine until the night before the famous first Moon walk was due to take place. There was a problem with the Parkes radio telescope in Australia. This was made all the worse when Armstrong and Aldrin, now in the lunar module, were flying down to the surface of the Moon. As Armstrong made the landing on the Moon with only 30 seconds of fuel left, it was scheduled for the two astronauts to have a sleep break. Armstrong understandably overruled this sleep break. They had just landed on the Moon, who would feel like sleeping? It would be like trying to keep children in bed on Christmas morning knowing all their presents are downstairs. The radio telescope at Parkes managed to find the signal just in time to watch Neil Armstrong descend the ladder and take his first step onto the Moon's surface.

1.8 Interferometry and Quasars

Also in the early 1960s, a series of experiments with the Mark 1 radio telescope at Jodrell Bank and other small telescopes at increasingly greater distances from the Mark 1 served to further develop the system of radio interferometry. Between the years 1960 and 1963, using the method of interferometry, a sky survey conducted by Cambridge University found that a number of the most powerful radio sources in the sky had quite small angular sizes. One particular object of interest was given the name of 3C273. The first "3" denotes that it was found during the third survey; the "C" stands for Cambridge and "273" denotes that this was the 273rd object found during the survey. Unfortunately the exact position couldn't be narrowed down enough to make an observation with an optical telescope. Armed with this information, the Parkes radio telescope in Australia, using occultations of the Moon, managed to narrow down the position of 3C273. This was done by watching the Moon transit across the sky and taking its exact position when the signal being received by the radio telescope was lost behind the Moon's disc. Since the Moon has an angular measurement of approximately half of a degree this was enough to narrow the search area so that the large optical telescope on Mount Palomar in California was able to see it optically.

This powerful radio source was given the name of a quasar. Quasar means "quasi-stellar radio source". Quasars emit energy over a wide range of wavelengths, including optical wavelengths, and they are thought to inhabit the centers of very remote galaxies. When quasar 3C273 was finally found optically its spectrum was found to be very unusual and unlike anything else ever seen before. It had very wide emission lines within its spectrum, which couldn't be identified, until an astronomer by the name of Marten Schmidt realized that the spectra was massively shifted towards the red end of the spectrum. When the redshift was worked out it was found to be 0.6. This meant that it was travelling at an extremely high recessional speed, far faster than anything recorded before, and therefore the object must be at a great distance from our galaxy.

In 1963 the first interstellar molecule was discovered by Sander Weinreb and a team of other radio astronomers, within the absorption spectrum of Cassiopeia A. This molecule was hydroxyl radical (HO). The hydroxyl radical molecule is the neutral form of the hydroxyl ion. Hydroxyl radicals are highly reactive and for this reason they are very short lived. If they were introduced into the Earth's atmosphere they would instantly react with it, but in space there is very little of anything apart from the odd gas molecule for the hydroxyl radical molecule to react with, therefore the hydroxyl radical molecule has a much longer life expectancy. The importance of this new discovery of the hydroxyl radical molecule in radio astronomy meant that, like the discovery of the 21 centimeters hydrogen line, it gave radio astronomers another wavelength in which they could probe into the interstellar clouds. Using the 18 centimeters (1,665.4018 megahertz) wavelength of the hydroxyl radical molecule meant that radio astronomers could see further into denser interstellar gas clouds than observing in the 21 centimeters hydrogen line. This was useful for exploring areas such as the Orion nebula that have active star formation regions, and other areas of interest around the expanding gases and material of supernova remnants. The discovery of the hydroxyl radical molecule also proved to astronomers that atoms could come together in the vacuum of space and form molecules. This led to more questions about the ability of more complex molecules to form and exist in the extreme cold vacuum of space.

1.9 Observation of Mercury

These new wavelengths opened up a new path to observe the planet Mercury, which has always been a problem for astronomers due to its small physical size of 4,880 kilometers (3,032 miles) in diameter. Mercury has an highly elliptical orbit which makes a huge change in the angular size of the planet as observed from the Earth's surface. The planet's proximately to the Sun also makes it difficult to observe. At greatest elongation Mercury is no more than 28 degrees away from the Sun as seen from the Earth. For this reason optical astronomers have never seen Mercury in a truly dark sky, but only in the twilight, be it either dawn or dusk, depending where Mercury is in its orbit. It is also dangerous to sweep the sky looking for Mercury as the Sun may accidentally be bought into view through either a pair of binoculars or a telescope, and this would instantly damage the observer's eyesight, potentially permanently. Damage could also occur to light sensitive equipment such as the chips used in CCD cameras.

When viewing the planet Mercury in a twilight sky it is always low down in the thickest part of the atmosphere where the "seeing" is at its worst. If lucky enough to catch a glimpse of the planet it appears to have phases similar to that of the Moon and has a pinkish hue to it. Surface detail is very unlikely to be seen with an amateur telescope, and even imaging of the planet through an amateur size telescope is no better, but high magnification images of the planet show that Mercury has a heavily cratered surface, like that of the Moon. All these problems in observing the planet with optical telescopes meant that the true rotational period of the planet was not known for sure, and could only be estimated. But, as most things in science and astronomy, the reality is far stranger and far more exciting than anyone would have imagined. The rotational period of Mercury was first thought to be the same as its orbit of 88 days, in the same way that the rotation of the Moon is roughly the same as its orbit around the Earth. This can be shown as a ratio of 1:1.

In 1965, using the 305 meter (1,000 feet) Arecibo radio telescope in Puerto Rico, radio astronomers Gordon Pettengill and Rolf Dyce carried out a number of experiments to bounce radar off the planet in order to find Mercury's true period of rotation. It was hoped that the returning echoes would prove or disprove once and for all the theory of the 88 day rotation of Mercury. It was found that the planet had a truly unique period of rotation and unlike any other planet within the solar system, and stranger than anyone could have ever thought. Mercury was found to have an orbital resonance ratio of 3:2. This means that the planet rotates three times on its axis for every two orbits the planet makes of the Sun. When this orbital resonance was calculated it was found that Mercury takes 59 Earth days to rotate once on its axis. This unique rotation coupled with its highly elliptical orbit, more eccentric than any other planet in the solar system, would make for some very interesting effects, if it was possible stand on the surface of Mercury. The Sun would be seen to rise, but as the Sun got higher in the sky it would appear to get smaller in size, due to the planet's highly elliptical orbit. Then the Sun would appear to backtrack across the sky again in the same direction from which it rose and set. Then it would rise again, but this time it would travel right across the sky with the same apparent change in size and set. This would be like seeing a double sunrise and one sunset, and would make for a totally unique and outrageous calendar.

1.10 The 'Big Bang' Theory and CMB

In the foment of all of the discoveries of the 1960s there were several theories doing the rounds that attempted to explain the origin of the universe. Some had an element of science to them, and others were rather esoteric, but there were two basic theories regarding how the universe came into being that most scientists believed could be true. The first was the "steady-state" theory and the second was the "big bang" theory. The "steady-state" theory was that the universe has always existed and will carry on existing forever without noticeable change. Stars will die but new stars will be born to take their place in an endless everlasting cycle. The "big bang" theory states that the universe had a beginning and it will have an end. The universe is said to have begun from a vast explosion from which all the matter in the universe came. According to the theory, this explosion was the start of everything including time and space, and is thought to have happened 13.75 billion years ago. The end of the universe could come in the form of a big crunch as gravity pulls everything back together when the expansion of the universe stops, or the universe could carry on expanding forever, ending its days as a cold dark place devoid of heat and light.

Fred Hoyle was a well respected British astronomer and a writer greatly in favor of the "steady state" theory. During a radio interview Hoyle was asked about his thoughts that the universe started with a large explosion. Hoyle told the interviewer his thoughts on the origin of the universe and how he preferred the steady state theory and made a comment about the theory that the universe started with an explosion as a "big bang" idea. This off the cuff comment stuck and was taken up and used by the other astronomers and physicist, such as Stephen Hawking, as the name to describe their opposing theory. There was some rivalry between the two theories, and both sides believed that their theory was correct. They agreed that a way should be found to prove which theory was correct once and for all.

While the battle of the "steady-state" and "big bang" theories raged on, two scientists, Robert Wilson and Arno Penzias, were working for Bell Laboratories in Holmdel, New Jersey. This was the same Bell Laboratories that Karl Jansky worked for when he found the radio source in the constellation of Sagittarius. Wilson and Penzias were using an interesting type of radio telescope, the 6 meter (20 feet) horn antenna. Its design was so unusual it was nicknamed the "sugar scoop" because of it having a striking resemblance to a huge scoop. Wilson and Penzias were originally going to use the horn antenna as a way of receiving the echoes from radio waves bounced off of balloon satellites. Balloon satellites were sometimes referred to as "satelloons". They were satellites that had been inflated with gas after being placed into orbit. But as soon as Wilson and Penzias switched on the receiver, they

1.10 The 'Big Bang' Theory and CMB

picked up all the usual sources of interference along with an annoying hissing sound that seemed to be always present. They really needed to know whether or not they were receiving the echoes that they wanted from the balloon satellites, so the two scientists had to painstakingly eliminate, or at least account for, every source of interference that their antenna and receiver was picking up. The two scientists set about eliminating each source in turn, but there was a source of interference, a steady hiss, that was always there and wouldn't go away.

Their first thought was that the hiss could be interference from the nearby city, but this was soon discounted as no matter what direction the antenna was pointing the hissing noise was always there. What was more puzzling was that it seemed to be at a constant level no matter which direction the antenna was pointing. This meant that the hissing noise wasn't coming from one particular source, and therefore it had to be something that was the same in every direction. This led Wilson and Penzias to think that the hiss could be coming from the receiver itself. Maybe this hissing noise was coming from a noisy power supply, or the electronics within the receiver producing thermal noise. So the two then tried using liquid helium to cool the receiver to −269 degrees Celsius (−452 degrees Fahrenheit), just 4 degrees above absolute zero, in order to remove all the thermal noise from the receiver circuitry itself. They did this in the hope that this would account for the problem, but the hissing was still there even at the extremely low temperatures. After the cooling of the receiver failed to remove the hissing, they turned their thoughts skyward in order to eliminate radio emissions from the galaxy, the Sun, and even the Earth, and still the hissing noise was present.

The two now took a less technical approach to remove the frustrating source of interference. A family of pigeons had decided it would be a good idea to make the inside of the horn antenna their home; and with birds comes the problem of their droppings and lots of them. Wilson and Penzias promptly frightened the family of birds away and took a broom to the inside of the horn antenna to clear it from the countless bird droppings. But no sooner had the two scientists frightened the birds away, then the pigeons came back again. This battle raged on for several weeks, and the hissing noise from the receiver was still there. It didn't even seem to matter whether the pigeons were there or not.

Wilson and Penzias discussed with other scientists their problem regarding the constant "hissing" noise coming from their receiver. One of these scientists was Bernard F. Burke. Burke told them about a paper that he had read about how if an explosion as great has the big bang had taken place then there must be a remnant or echo to it, suggesting that this echo may possibly be found within the wavelengths they had been using. The pair then realized that they had accidentally found the faint echo of the largest explosion in history, that of the big bang. This then gave the big bang theorists all the proof they needed and the steady-state theory soon fell out of favor, though it is understood that Fred Hoyle refused to believe it. Robert Wilson and Arno Penzias were awarded the Noble Prize for Physics in 1978 for the discovery of the Cosmic Microwave Background (CMB) radiation. This hissing noise is all that is left of the big bang, a faint ghost from the start of the universe. The CMB radiation can easily be picked up on a television set or a radio by tuning a television set or radio to an unused channel. Approximately 1 % of the white noise or static that can heard is the Cosmic Microwave Background radiation from the big bang.

1.11 Neutron Stars and Pulsars

CMB was not the only startling discovery to come out of radio astronomy research. In 1965 in an unused field outside the city of Cambridge a radio telescope was going to be built that would in 1967 discover one of the strangest objects ever discovered in the universe. The quest to understand this discovery would push the laws of physics to breaking point. Antony Hewish and a group of research students, one of whom was Jocelyn Bell, wanted to build a radio telescope to carry out radio observations of the sky, but as always funding for such a telescope wasn't very forthcoming. Hewish, with the budget that he had, settled on a multiple di-pole array for the antenna for the radio telescope. To save money it had to be built by Hewish and his group of students. This work consisted of hammering a large number of wooden stakes into the ground and stretching wires between the wooden stakes to form the antenna. Hewish and the research students took turns hammering the wooden stakes into the ground and fastening the wires to them. Then all the wires had to be soldered together to complete the circuit and turn this jumble of wires into an antenna.

After 2 years of work they had built an odd-looking radio telescope made up of wooden stakes and a jumble of wires that crisscrossed the field. The antenna covered an area of approximately 16,200 square meters (19,360 square yards). After its completion a research student by the name of Jocelyn Bell was chosen to take charge of the running of the receiver and chart recorder. At a later date, in an interview, she recalled how the chart recorder was producing 29 meters (96 feet) per day of recording chart and that she had to examine all of it. Within months of the radio telescope being switched on Bell had generated several miles of chart recording paper. On examination of the chart paper she noticed at a certain time of day a signal was received and recorded on the chart paper, a signal that repeated every one and one third seconds as regular as clockwork. This pattern was so regular that it crossed her mind that it could in fact be an artificial signal, possibly from an extraterrestrial civilization, and for this reason Bell gave the signals the nickname LGMs, short for Little Green Men.

The next thing that had to be done was to find out exactly where the LGMs signal was coming from, but as mentioned above radio telescopes are notorious for having very poor resolution qualities. This was done in the same way Jansky did it by finding out the exact part of the sky that was travelling through the antenna beam of the radio telescope at the time the signal was being received and at its strongest. The width of the antenna beam with this type of design would have been quite large so only an approximation of where the signal was coming from could be made. All manner of theories were thrown into the hat regarding what this signal that repeated

1.11 Neutron Stars and Pulsars

every one and one third seconds could be. Hewish theorized that the object must be small in physical size or it would have been already observed, but it also must be very powerful to generate such a signal. Hewish made enquires and asked a number of theoretical physicists if any had theories regarding what this very small and very powerful object could be. The answer that kept coming up was a neutron star. Up to this point in time neutron stars had only existed within the minds of the theoretical physicists and there was no physical evidence that they existed or in fact that they could exist.

A neutron star is an extremely small and very dense object, so dense that it is thought that a sugar cube sized piece of matter from a neutron star would have a mass of 100 million tons. Neutron stars are formed when a massive star reaches the end of its life and explodes as a supernova. As this outward explosion takes place the star blasts off its exterior layers. The outward explosion masks an implosion that compresses the leftover core. The compression forces are so high within the core that the empty space that is usually present in all atoms between the orbiting electrons and the atoms nucleus is lost. The electrons and protons are merged into neutrons (hence the name neutron star). All the neutrons are forced together so there is almost no room between each neutron, accounting for the very high density of a neutron star.

It is thought that a neutron star that originally had the same mass of the Sun would be compressed down to a diameter of around 20 kilometers (12 miles). Stranger still, the surface of a neutron star is thought to be a form of iron that is 10,000 times as dense as the iron found on Earth. Below this crust of super-dense iron is thought to be a mantel of liquid neutrons. What lies at the core of a neutron star is still under debate because it is so dense within the core that the laws governing physics start to break down. Some theorists believe it could be a new exotic form of neutrons, yet to be classified.

Neutron stars also rotate very quickly on their axis. A better word to describe this very quick rotation would be "spin". The spin of a neutron star can be many times per second. This is due to the conservation of angular momentum law, in the same way that an ice skater on the ice will spin faster if they bring their arms in closer to their body. The gravity on a neutron star is immense. By way of an illustration, if the same gravitational force of a neutron star was applied to the Earth and it's atmosphere the top of the atmosphere would be less than 6 millimeters (0.25 inch) from the Earth's surface.

As well as having a huge gravitational force, any magnetic field that the star originally had would now be concentrated within this small fast-spinning neutron star. Concentrating this magnetic field produces two high-powered beams of energy that come out of the magnetic poles of the neutron star, in the same sort of way that the beams of a lighthouse sweep out to sea. If one of these beams from the neutron star then sweeps across a radio telescope it will produce a signal.

It was found that the signal from the radio telescope coincided with a part of the sky that contained the constellation of Taurus, but this is as close as could be estimated. Optical astronomers were then asked if there was anything in that particular part of the sky that warranted further investigation, and the answer was: the supernova remnant that is the Crab nebula. Optical astronomers already knew about the Crab nebula being a supernova remnant. As the cloud of gas of the nebula was still expanding, they even had a date at which the supernova explosion took place: the year 1054 AD. This information had been found in an ancient Chinese manuscript dated from that year and containing descriptions of the observations made by Chinese astronomers who recorded the appearance of a "guest star" that was observed in the same part of the sky where the Crab nebula is now. It shone for 22 months before fading. This guest star could be seen in daylight like the planet Venus. For a time it was also possible to write by its light at night. This proved the theory behind the supernova, and armed with this new information optical astronomers observed the Crab nebula again, and with the use of high-speed photography. It was found that the central star was the corpse of the long dead star spinning at 30 times per second. This was what Bell had observed on the chart recording from the radio telescope.

Up to this time of 1967 this was the most powerful thing ever photographed in the universe. These strange objects were given the name "pulsars" because of the pulses that they produced on a chart recorder. Antony Hewish and Martin Ryle both shared the 1974 Nobel Prize for Physics for the discovery of the first pulsar. Ryle was chosen to share the prize. As in the early 1960s Hewish and Ryle had developed the technique of aperture synthesis which is a form of interferometry. There was quite a bit of controversy over Hewish sharing the Noble Prize with Ryle for the discovery of the first pulsar. It was argued that Jocelyn Bell should have been credited with her part in the discovery. After all, it was Bell who had found the original signal and had realized its importance and brought it to the attention of Hewish. Even Fred Hoyle, the astronomer who favored the steady-state theory of the universe, added his support for Bell to receive proper recognition. Hoyle argued that she should be awarded co-discoverer of the pulsar and therefore the Prize should be shared with her. Unfortunately Bell was not given the credit that she deserved.

If wishing to hear an audio clip from a pulsar visit www.astrosurf.com/luxorion/audiofiles-pulsar.htm. Once there, click on an icon to hear a short audio clip of a number of different pulsars. To the right of the icon is a brief description of the pulsar and its origin. The audio sound between different pulsars is quite marked. Some sound like the speeded up ticking of a clock, but the millisecond pulsars sounds like a hammer drill boring a hole in a wall and is quite ear-piercing.

1.12 Arecibo Transmission and 'Wow!' Signal

Some of these sounds may have been picked up by the giant Arecibo radio telescope in Puerto Rico, an extremely sensitive radio receiver, so sensitive that it could pick up the signal from a cell phone if it was used on the planet Jupiter. In 1974 the giant dish was going to transmit the most powerful transmission ever made in history, which was going to be deliberately beamed into space. This signal was meant

1.12 Arecibo Transmission and 'Wow!' Signal

to broadcast a great cosmic hello to let any extraterrestrial civilizations within reception of the transmission know that the planet Earth was here. This idea was met with mixed reactions. Some thought it was a great concept, while others had misgivings about the project. The main cause for concern was what if an extraterrestrial civilization that wasn't particularly friendly received the transmission, and they were intent on world domination, the Earth has just screamed out "we're over here". Some believed that it would be a much better plan to just eavesdrop on any cosmic communications first, in order to see if any potential extraterrestrial civilizations was friendly or not. It was pointed out to anyone who opposed the transmission that it would take thousands of years to arrive at another planet, so it could be expected to take just as long to receive a reply. Also it was hoped that if an extraterrestrial civilization had the technology to travel the thousands of light years to get to the Earth, they would be of good character and wouldn't want to take over the planet Earth.

The transmission was due to take place from Arecibo on November 16th 1974. It was part of the celebrations to mark a major upgrade to the radio telescope. The signal was going to be sent towards the globular star cluster known as M13 in the constellation of Hercules, named after a hero in Greek mythology. The star cluster M13 is the best globular star cluster in the northern hemisphere, visible as a faint misty patch to the unaided eye but beautiful viewed by telescope. This cluster is thought to contain around 300,000 stars or more, and is about 25,000 light years away. It was chosen as a target because of the great volume of stars within the cluster: with the greater number of stars it was hoped that there would be a greater chance of finding a planet orbiting one of these stars. With the upgrade of the Arecibo radio telescope, its new more powerful transmitter now measured in megawatts. The 305 meter (1,000 feet) dish was capable of beaming out a transmission to a relatively small area of the sky. This meant that all of the power could be concentrated into an outgoing signal equivalent to many megawatts of power.

The target had been chosen and the most powerful transmitter on the Earth was ready, but what was going to be broadcast? A message needed to be written. That if received and decoded, would prove to an extraterrestrial civilization, that there was other life in the universe. Frank Drake of Search for Extra-Terrestrial Intelligence (SETI), wrote a draft of what he thought should be in the message. With the help of Carl Sagan (who would be involved at a later date with the gold discs on the two Voyager space probes) and aided by other astronomers and mathematicians who pooled their knowledge, a special mathematical message was written. The message was to be broadcast in the most basic fashion possible, that of binary code "ones and zeros", and would be made up of 1,679 bits. This number was chosen because it is the product of two prime numbers, namely 73 and 23. This meant the message could either be written as 73 columns and 23 rows or 73 rows and 23 columns. This was deliberately chosen because if the message was received by an extraterrestrial civilization, it was hoped that by using mathematics the message could be reassembled in one of two ways. If the wrong way was chosen, namely using 73 columns and 23 rows, the message would look like a random pattern, if on the other hand the right way was used, 73 rows and 23 columns, an image would be formed that couldn't be mistaken for anything other than a message from another civilization.

The message contained images of a human being, a radio telescope dish, and a double helix to represent DNA. The numbers 1–10 were also shown, along with a representation of the solar system and other information such as physical and biological information from the Earth. Information about chemical elements was also included within the message. All these images were not very detailed and were basic in appearance. They were made up of large square pixels similar to the very first video arcade games back in the 1970s. The entire message was broadcast in less than 3 minutes and, travelling at the speed of light towards the star cluster M13, is due to arrive at its destination in just under 25,000 years time. If any civilization does receive and decipher the message and then decides to send a reply, the response will take another 25,000 years to get back to the Earth. Additionally, as everything in space is moving, and in some cases very quickly, the star cluster M13 will have moved its position relative to the Earth by the time the transmission gets there and vice versa. The solar system and the Earth which it contains will have moved part way around on its 220 million year orbit of the galactic center in the 50,000 years it would take to get a reply, possibly rendering the whole exercise just academic anyway.

At the same time the Ohio State University Radio Observatory in the United States began to conduct what would be the longest running SETI program, in existence for 22 years from 1973 to 1995. A radio telescope at OSU was constructed in the late 1950s and finished in 1963. Located in Delaware, Ohio, the telescope was built to survey the sky for wideband extraterrestrial radio sources and was given the name of "Big Ear". This name must be without a doubt the most appropriate name possible for a radio telescope. The Big Ear telescope was in a fixed position and relied on the rotation of the Earth in order for it to scan the sky. This meant that it could monitor a point in the sky for approximately 70 seconds before the rotation of the Earth moved the object out of the telescopes beam.

The morning of August 15th 1977 would start the same as every morning for radio astronomer Jerry R. Ehman. Little did he know that while using the Big Ear radio telescope he would receive a signal from space that to this day remains a total mystery. This signal became known as the "Wow!" signal. What made the "Wow!" signal so special was the bandwidth at which the signal was received. Natural radio emissions are usually of a wide bandwidth in nature and can be received at a number of different frequencies simultaneously. A manufactured signal, such as one produced by radio transmitters on the Earth and used to transmit a local radio station, is what is known as a narrow band signal. This is what allows the listener to tune in to a number of different stations clearly without any overlap in frequency between each station. This gave the impression that wherever and whatever transmitted the Wow! signal must have been manufactured in order for the signal to have such a narrow bandwidth.

When the "Wow!" signal was received it was found that it was at a much higher level than the background noise of space, something approaching 30 times the level

1.12 Arecibo Transmission and 'Wow!' Signal

normally perceived as the average level for the background of space. This was beginning to look more and more like a deliberate transmission. The next interesting thing was the frequency at which the signal was received. When Frank Drake carried out the first SETI search back in 1960s, he chose to listen for extraterrestrial transmissions at the 21 centimeters hydrogen line at the frequency of 1,420 megahertz. He chose this because hydrogen is the most abundant substance in the universe. The Wow! signal had a narrow band width very closely matching the frequency of 1,420 megahertz of the 21 centimeters hydrogen line. All these factors seemed to be too much of a coincidence for anything that could have been produced naturally. The signal was received for approximately 70 seconds. That coincided exactly with how long the signal was within the beam of the radio telescope, before the Earth's rotation moved the telescope and the signal was lost. On the original printout Ehman wrote the now famous "Wow!" at the side of the signal. Investigations were carried out later to prove that this signal did in fact originate from deep space and was not anything within the solar system, also that it was nothing to do with any manmade source or anything that the military were involved with. Any manmade source of interference at this frequency would have been highly unlikely as this particular frequency is one of a number of protected frequencies which are not used because of their significance to radio astronomy. The part of the sky that the signal was coming from was within the constellation of Sagittarius. Ehman made several further attempts over the next few weeks and months to try and receive the signal again, but to no avail.

Several other radio astronomers have tried over the years to receive the signal using more modern and more sensitive radio telescopes, but every attempt to date has failed, and exactly what the signal was and the story of its origin still remains a mystery. All that is known with any certainty is that it was extraterrestrial and it came from somewhere within the direction of the constellation of Sagittarius. There is a tradition amongst SETI observers. They always have a bottle of Champagne chilling in a refrigerator at the radio telescope observatory where they are carrying out their observations, just in case they do receive an extraterrestrial signal. Whether this tradition came about because of the Wow! signal is not known, but Jerry Ehman could be forgiven if he opened a bottle, considering all the factors involved.

SETI have produced a protocol that they recommend should be followed, just in case a proven extraterrestrial signal is received. The protocol goes something like this: on receiving an extraterrestrial signal it must be confirmed by another radio telescope installation; once proven authentic, the heads of state of the country that first received the signal should be told of its existence; next, the heads of state from that country should inform the United Nations of the signal in order that a decision can be made about how to proceed, for example whether or not to reply, and if the decision was made to reply what would the reply say? Such a decision will probably have to be taken with the advice of scientists and philosophers. It would also depend on where the signal originated and the distances and timescales involved as any potential conversation with an extraterrestrial civilization could theoretically span many generations. So the children of future generations will need to be told of the existence of the signal and roughly a timescale in which to listen out for a reply. It would be awful to think that the radio telescope that had waited so many years for a reply was undergoing maintenance and we missed the reply.

As can be seen from this short introduction to radio astronomy, it has been and still remains a useful tool for making observations of objects that could not otherwise be observed within optical wavelengths. Radio astronomy has proven its worth in a very short time by demonstrating the existence of some truly remarkable and exotic objects within our galaxy and beyond. The next step is to understand its mechanics better.

Chapter 2

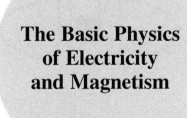

The Basic Physics of Electricity and Magnetism

All matter in the universe, the air we breathe, the paper pages of this book, are all made up of atoms. Take for example a copper wire, since most electrical cables are made of copper. If a piece of copper wire were cut into smaller and smaller pieces we would eventually get down to the size of an atom. An atom of copper would be the smallest part of material that could exist and still keep its identity as copper.

2.1 Building Blocks

All atoms, with the exception of the hydrogen atom which has no neutron, are made up of three basic parts, neutron(s), proton(s) and electron(s). The nucleus of the atom is made up of the proton(s) and neutron(s) with the electron(s) orbiting round the nucleus. The structure of the atom is sometimes likened to that of the solar system, where the nucleus would be represented by the Sun and the planets in orbit around it would be the electrons. This is not strictly true, as the planets orbit the Sun in the same plane, while the electrons that orbit an atom do so in all different planes.

The Fig. 2.1 is of a helium atom. It has a nucleus which is made up of two protons, and two neutrons, and it also has two orbiting electrons. The rest of the volume between the nucleus and the electrons is just empty space. To give an idea of how much empty space is contained within an atom, if an electron were the size of a table tennis ball, the innermost electron(s) orbit would be in the region of 1.6 kilometers (1 mile) from the nucleus of the atom.

The proton carries a positive electrical charge, the neutron carries no electrical charge, and the electron carries a negative electrical charge. The protons and

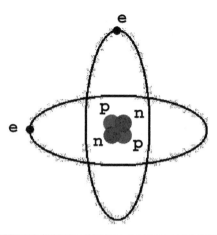

Fig. 2.1 Helium atom (not to scale)

electrons cancel each other out, and therefore the atom is electrically neutral. As a rule, an atom has the same number of protons as electrons, and this leaves the atom electrically neutral.

If an electron is lost from an atom then the overall charge of the atom will be changed to positive. This is known as a positive ion. If an electron is added, the overall charge will change to negative. This is known as a negative ion. This can happen if the atoms are exposed to higher energy radiation, for example ultraviolet light from the Sun. This process is known as ionization.

Other materials have atoms with electrons in orbits at greater distances from their nucleus, in the same way that Jupiter has a greater orbit than Venus from the Sun. The outermost electrons are not as tightly bound to the nucleus as the electrons on the inner orbits of an atom. These outer electrons in some materials will swap orbits with their neighboring atoms, and these are known as free electrons. Some materials, where there isn't an abundance of free electrons and where their electrons are more tightly bound to their nucleus, are known as insolating materials, for example rubber, glass, and air. Although if the electrical pressure is great enough, as in the case of a lightning bolt, then even air can be made to conduct electricity. So there are no perfect insulators. In other materials, such as copper and gold, there are lots of free electrons that are relatively free to move randomly and swap orbits with electrons of other neighboring atoms. If an electrical charge is applied to the material, all of the free electrons can be made to move in the same direction by this electrical pressure. These are known as conductors, such as copper wires.

By way of illustration, imagine that a short length of garden hose was used to represent a conductor, such as a piece of copper wire. Imagine the hose were to be filled with marbles to represent the free electrons already in the copper wire. If a marble was pushed in one end, then a marble would fall out the other end. If ten marbles were pushed in, then ten marbles would come out the other end. The flow of marbles would be the equivalent to a flow of electrons through a conductor.

This flow of electrons is called electrical current and is measured in amps. The force or electrical pressure used to push the electrons along a conductor, is voltage and is measured in volts. There is another force that needs to be taken into consideration, that of resistance. A copper wire isn't a perfect conductor. All conductors will cause a small amount of electrical energy to be lost within the conductor. This is known as resistance, and is measured in Ohms (Ω). (There are superconductors where the resistance can be almost removed completely, but these need to be cooled to a few degrees above absolute zero). If the garden hose example is used again, but with water this time, the amount of water coming out of the hose ("the flow") will equate to the electrical current, and the water pressure forcing the water through the hose would be the voltage. If somebody stood on the hose and restricted the flow of water, this would be equivalent to electrical resistance.

All three of the above, current, voltage, and resistance are all connected mathematically by Ohm's Law. $V = I \times R$

where

V = voltage in volts (V)

I = current in amps (A)

R = resistance in Ohms (Ω).

Example: It would take a voltage of 1 volt to cause a current of 1 amp to flow through a resistance of 1 Ohm.

2.2 Magnetism and Wave Properties

Magnetic fields and electric currents are different, but they are interconnected. When an electric current flows through a conductor, such as a wire, a magnetic field is generated around the conductor. If the wire is coiled then this can concentrate the magnetic field around the wire. Please see Fig. 2.2.

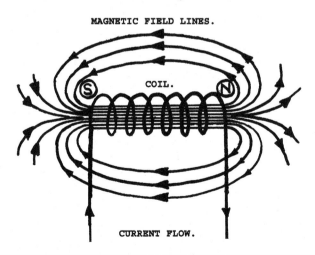

Fig. 2.2 Magnetic field around a coiled wire

If the wire is coiled around a piece of iron and connected to a battery so that a current can flow through the coil, the iron will become magnetized. The magnetic field generated by the electrical current flowing through the wire causes all the small magnetic crystals contained within the iron to line up in the same direction, with north at one end and south at the other, and produces an electromagnet, but when the electrical current is switched off the alignment of the magnetic crystals within the iron is lost and so is the magnetism.

If, on the other hand, the same coil of wire had the iron removed and instead of connecting the coil to a battery the coil was connected to an ammeter (a device for measuring electrical current) and a bar magnet was passed in and out of the coil, an electrical current will be generated within the coil of wire. If an object is placed within a magnetic field the object will be affected. For example, if iron filings are placed around a bar magnet they will line up with the magnetic field lines, or in the case of a compass the needle will be affected. If the field lines are followed it will be found that they converge at the poles of the magnet.

The Earth has a magnetic field thanks to the molten core. This magnetic field permeates into space and protects the Earth from the charged particles within the solar wind by deflecting them along the magnetic field lines (this will be covered in detail in a later chapter). So, it can be seen that the flow of electrical current can produce magnetism, also magnetism can be used to produce an electrical current.

Electromagnetic Waves

If a stone is thrown into a calm pond the ripples that emanate from the point where the stone hit the water are a form of wave. Electromagnetic waves were once a bit of a mystery. Scientists just couldn't understand how electromagnetic waves travelled, especially in a vacuum. Electricity is the flow of electrons, and this flow requires a conductor such as a wire to travel through. So scientists thought there must be an unknown substance, a medium yet to be discovered, which the electromagnetic waves travelled through. Like the stone thrown into the pond, the wave travels through the water and uses the water as a medium to travel. Scientists even gave this unknown medium a name Luminiferous aether. This mysterious Luminiferous aether was thought to be an unseen substance that permeated through space and in fact right through the universe, and allowed the electromagnetic waves of light and heat from the Sun to reach the Earth.

Scientists tried for many years to find the mysterious Luminiferous aether and failed, and thanks to Guglielmo Marconi's Trans-Atlantic radio transmission in 1900 the Luminiferous aether was finally proven not to exist. An experiment used by science teachers in schools demonstrates this very well. The experiment consists of an electric bell inside a large glass jar that is sealed and air tight. The teacher starts the bell ringing and slowly pumps out the air until the bell can't be heard, but the bell is still visible. This proves that sound waves need a medium in which to travel "the air", and what we hear as sound is nothing more than compression waves carried in the air itself. No air meant that there were no air molecules to vibrate and

2.2 Magnetism and Wave Properties

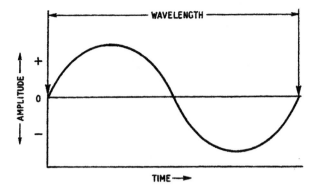

Fig. 2.3 Image showing wavelength and amplitude of a sine wave

carry the sound. But on the other hand the bell can still be seen, so this proved that light, an electromagnetic wave, can travel unhindered in a vacuum. Therefore, as we can see the Sun and feel its heat, it can be concluded that electromagnetic waves can travel the 150 million kilometers (93 million miles) from the Sun to the Earth, travelling through the vacuum of space without the need for a medium in which to travel.

Wavelength and Amplitude

Wavelength is a measure of the distance between the wave patterns repeating themselves. Please see the Fig. 2.3.

This is the image of a sine wave, the type of wave is not important as this is the same for all waves, square, triangle, and so on. The graph is of time across the bottom axis, wavelength across the top axis and amplitude on the vertical axis. It can be seen from the graph that this shows one complete oscillation of the wave before it starts to repeat itself, and this is known as the wavelength of the wave. Wavelengths can be measured in millionths of a millimeter or thousands of kilometers.

Amplitude is a measure of the signal strength of the wave, and the higher the peak to peak distance the more power or amplitude is in the signal. Using the stone thrown in the pond example again, if a small stone is thrown a small amplitude wave will result, if a large stone is thrown a large amplitude wave will be produced.

Frequency

All electromagnetic waves travel at the speed of light, which is 300,000 kilometers (186,200 miles) per second in a vacuum. The key word here is "vacuum". As electromagnetic waves can slow down, depending on the medium through which they are moving, for example glass, water, etc. electromagnetic waves never travel faster than the speed of light, as this would break the fundamental laws of physics that nothing can travel faster that the speed of light, except in the movies. So, for the purpose of this book, we will assume that electromagnetic waves always travel at the speed of light.

Frequency is the measure of how many repeat patterns of the wave travel past a point per second. For instance, if we use the above image and say the time line across the bottom is equal to one second, then this means the frequency of the wave would be 1 hertz (abbreviated as 1 Hz). If the pattern of the wave is repeated twice a second we would say the frequency was 2 hertz (2 Hz), and so on. If we look at the radio in a car and the frequency marked on the slider for example, a frequency of 100 MHz (megahertz), this indicates that the number of repeat waves per second would be 100 million. It is also worth mentioning that some older radio astronomy books refer to the frequency in Cycles per second or C/S. This is exactly the same as hertz, so 20 hertz is the same as 20 C/S and 20 megahertz is the same as 20 MC/S.

The Relationship Between Wavelength and Frequency

If an antenna for a particular frequency is to be made, for example the antennas for the Radio Jove receiver, then the wavelength must be known so the antenna can be cut to size (this will be covered later). There is a simple equation that can be used to convert frequency to wavelength and vice versa.

$$Wavelength = \frac{300,000}{Frequency}\ (the\ speed\ of\ light).$$

$$Frequency = \frac{300,000}{Wavelength}.$$

Polarization of Waves

An electromagnetic wave is made up of two elements, hence the name comprising "electro" and "magnetic". The electrical part of the wave is not the type of electricity we are familiar with, which is the flow of electrons through a conductor such as a wire. But electromagnetic waves do carry the energy of electricity and magnetism, and if they come into contact with a conductor, for example an antenna, they would induce an electrical current to flow through that conductor or antenna. Although this current flow would be extremely small it would be enough for a suitable receiver to detect it and amplify this into a useable signal. The electrical and magnetic parts of an electromagnetic wave travel in the same direction, but are offset to each other by 90 degrees. Please see image below.

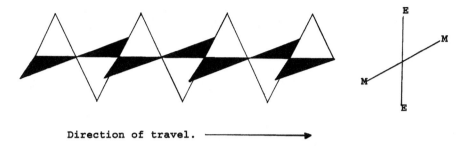

Fig. 2.4 Polarization of electromagnetic waves. An example of vertically polarized waves. Horizontal polarized waves will have the force lines the opposite way round

Polarization deals only with the electrical energy within the wave. It is important to know the polarization of the incoming electromagnetic waves in order to maximize the ability of the antenna to receive them. If it were possible to see these waves approaching, and if the movement of the electrical energy of the wave was moving up and down, the magnetic energy would be moving left to right, meaning this wave would be classed as being "vertically polarized". On the other hand, if the electrical energy of the wave was moving left to right and the magnetic energy was moving up and down, then this wave will be "horizontally polarized" (Fig. 2.4).

The polarization of an electromagnetic wave may not be fixed, and can rotate continually, sometimes randomly, either clockwise or counter clockwise. This is known as circular polarization. This is especially so if the wave has encountered anything that has caused it to change direction, for example reflection from an ionized meteor trail.

2.3 The Electromagnetic Spectrum

We are taught from an early age that the color red is associated with hot things also danger, such as stop lights on traffic signals, warning signs, etc. The color blue, on the other hand, is associated with cold. Someone may even say: "I'm so cold I've turned blue". A good example is a washing sink, the red tap is for hot water and the blue tap is for cold. But in nature the opposite applies. For example, red stars are cool, with surface temperatures around 3,000 Kelvin, while blue stars are hot, with surface temperatures around 20,000 Kelvin. The blue end of the electromagnetic spectrum carries far more energy than the red end. These energy levels are measured in electronvolts (eV).

An electronvolt is the amount of kinetic energy that a particle or wavelength carries.

For example, at one end of the electromagnetic spectrum there are gamma rays with energy levels of 124 kiloelectron volt, and at the opposite end of the electromagnetic spectrum we have the sub-hertz range of radio signals with energy levels as low as 12.4 feV. The more energy that a wavelength carries the more dangerous it is, especially to living tissue and the cells it contains.

Gamma Rays

Gamma rays are the most energetic wavelengths. They have wavelengths of 10 picometres and shorter. These are blocked by the Earth's atmosphere, which is good as they carry so much energy that they can smash into living cells and damage the nucleus within the cell and change its DNA. (Remember what happened to Dr. Banner, gamma rays turned him into the Incredible Hulk).

Gamma rays can damage a cell's DNA. This can cause it to mutate, this mutation can become cancerous. Gamma rays and other high energy wavelengths can be a real problem to any future astronauts flying to the Moon. They are thought to be emitted by high energy pulsars and supernovae explosions. Also a strange phenomenon known as gamma ray bursts have been recorded. These are short duration bursts of gamma rays which seem to come from a point source in space and are still being researched to find out where and what they are. Several theories have been put forward such as hypernovae (a super-supernova), but as yet they remain a mystery.

X-rays

X-rays are divided into two types, hard X-rays and soft X-rays. Hard X-rays have a wavelength of 10–100 picometres and soft X-rays have wavelengths of 100 picometres–10 nanometers. The longer wavelengths of soft X-rays are the type used by hospitals to find broken bones. These can also be used to cure or prolong the life of some cancer patients by irradiating their tumorous cancers. The Earth's atmosphere is opaque to this wavelength.

X-ray astronomy began in the late 1940s with the discovery that the Sun emits X-rays.

This particular wavelength has been proven to be a great aid in viewing the Sun, as it shows details that are masked in other wavelengths such as visible light and hydrogen alpha. X-rays are also a useful wavelength for looking at the Moon. The Moon itself doesn't emit X-rays, but when an X-ray flare is emitted from the Sun it travels through the solar system and when it hits the Moon it can cause certain elements within the Moon's surface to glow at X-ray wavelengths. This occurs in a similar way to how the color white seems to glow under an ultraviolet light source. This allows astronomers to study the properties of these elements on the lunar surface, and see detail that is not possible at visible wavelengths.

X-rays are also thought to be emitted by black holes. As gases are drawn into orbit around a black hole the speed at which the gases orbit gets faster and faster. This causes the gases to become superheated to the point where there is that much energy within the gas molecules that they emit X-rays.

2.3 The Electromagnetic Spectrum

Ultraviolet

The ultraviolet part of the spectrum can be sub-divided into two categories, extreme ultraviolet, with wavelengths of 10–100 nanometers, and near ultraviolet, with wavelengths of 100–350 nanometers. This near ultraviolet wavelength is the one responsible for suntans. Ultraviolet has proven very useful in the study of white dwarfs, a corpse of very dense material remaining after a star with a similar mass to the Sun has ended its life. Ultraviolet has been used to study the atmospheres of planets, such as Jupiter. Ultraviolet was one of many wavelengths used to study Jupiter's atmosphere after the comet Shoemaker – Levy 9 slammed into Jupiter's atmosphere. Ultraviolet has been used to study details in the upper cloud layers of the planet Venus and Saturn's planetary ring system. Hot blue stars also radiate strongly in ultraviolet light. An example of this is M45 the Pleiades, a beautiful open star cluster. Another very important part of ultraviolet radiation is the effect that it has on the ionosphere and in turn the effect on the radio astronomy projects discussed in later chapters.

Visible Light

We are all familiar with visible light. Its wavelength is approximately 400–700 nanometers. As a percentage the visible light portion of the electromagnetic spectrum is very small. This is where the Earth's atmosphere starts to become transparent. At 400 nanometers light is violet, the amount which can be seen at this wavelength depends on the individual. For example, it has been found people are more sensitive to the violet end of the visible light spectrum after a recent cataract operation.

The color blue comes in at around 450 nanometers, followed by green at 500 nanometers, yellow at 550 nanometers, orange at 600 nanometers and finally red at 700 nanometers. Electromagnetic radiation has a particularly unusual property about it. If we consider visible light, as we are so familiar with it, it has some of the properties of both a wave and a particle, but not all the qualities of either one. Experiments have been carried out using thin slits cut into a screen that produces a diffraction pattern on a nearby wall, but the diffraction patterns that are produced never seem to completely fit what scientists are expecting to see, and thus proving once and for all this unique property, and this still remains a mystery.

Infrared

The infrared part of the electromagnet spectrum covers quite a large swath, including all wavelengths from 750 nanometers to 1 millimeter. Infrared is sub-divided into three groups, near-infrared at a wavelength of 750–2,500 nanometers, mid-infrared at 2.5–10 micro meters, and far-infrared at 10 micro meters–1 millimeter.

The Earth's atmosphere is opaque to much of this wavelength and water vapor absorbs quite a lot of it. A limited amount of infrared astronomy can be done from the Earth's surface, but this has to be done from high mountains and the environment must be dry, as water vapor in the Earth's atmosphere will absorb some infrared wavelengths. There are airborne infrared observatories which are capable of flying above 99 % of the water vapor within the Earth's atmosphere. Infrared astronomy is extremely useful for looking through dust clouds that are opaque to visible light, for example through the plane of the galaxy towards the galactic center, and for observing star birth in such as the Orion nebula.

Submillimeter

The submillimeter wavelength is 0.3–1 millimeter. This part of the electromagnetic spectrum is where optical detection methods and radio detection methods meet. The antennas must be made to a higher degree of accuracy than other radio antennas operating at longer wavelengths. These telescopes must be placed at high altitude on the top of mountains, to get through the thickest part of the atmosphere. The environment must be very dry, as water vapor in the atmosphere can also block this wavelength. Areas such as the Atacama Desert in Chile are suitable. Here a large submillimeter array is being built on the top of a high altitude plateau. The radio telescopes are built at lower altitudes and then are moved up the mountain by a specially engineered transport vehicle. Once the radio telescopes reach the plateau they are then fitted into position. At this altitude construction workers have to work using breathing apparatus as the air is noticeably thinner. This is a new area of research in astronomy, and the aim of astronomers using this wavelength is to better understand molecular clouds, in order to better understand star formation, and the evolution and formation of galaxies.

2.4 Radio Frequencies and Characteristics

Now we reach the main concern of radio astronomy – the radio part of the electromagnetic spectrum. It is characterized by very large wavelengths, ranging from 1 millimeter to 100,000 kilometers or more. Unlike the other sections of the electromagnetic spectrum, where the frequency at which one section changes to another can sometimes be a little vague (for example, near infrared and mid-infrared, and also where hard X-rays meet soft X-rays), the radio spectrum is divided into the designated frequencies described below. Although even these can get grouped together, for example very low frequency (VLF) and the four groups below it are sometimes treated as a large sub group and referred to collectively as very low frequency (VLF).

Name: Terahertz
Abbreviation: THz
Frequency: 300–3,000 gigahertz
Wavelength: 1 millimeter–100 micro meters

2.4 Radio Frequencies and Characteristics

Examples: This is still experimental, but could prove a replacement for X-ray imaging where normal X-ray equipment could not be used because of access or because the area of the body on which it is going to be used could be damaged by X-rays. It also has scientific uses, such as higher frequency processors to enable faster computing and the relatively new field of submillimeter astronomy.

Name: Extremely high frequency
Abbreviation: EHF
Frequency: 30–300 gigahertz
Wavelength: 10–1 millimeters
Examples: Microwave remote sensing and microwave radio relays.

Name: Super high frequency
Abbreviation: SHF
Frequency: 3–30 gigahertz
Wavelength: 100–10 millimeters
Examples: Modern radar, microwave devices.

Name: Ultra high frequency
Abbreviation: UHF
Frequency: 300–3,000 megahertz
Wavelength: 1 meter–100 millimeters
Examples: Television broadcasts, cell phones, GPS, Bluetooth devices.

Name: Very high frequency
Abbreviation: VHF
Frequency: 30–300 megahertz
Wavelength: 10–1 meters
Examples: FM radio, ground to aircraft and aircraft to aircraft communication, maritime communication.

Name: High frequency
Abbreviation: HF
Frequency: 3–30 megahertz
Wavelength: 100–10 meters
Examples: Shortwave broadcasts, amateur radio and other commercial radio broadcasting and communication.

Name: Medium frequency
Abbreviation: MF
Frequency: 300–3,000 kilohertz
Wavelength: 1 kilometer–100 meters
Examples: Medium wave radio broadcasting.

Name: Low frequency
Abbreviation: LF
Frequency: 30–300 kilohertz
Wavelength: 10–1 kilometers
Examples: Long wave radio broadcasting, Navigation and timing signals.

Name: Very low frequency
Abbreviation: VLF
Frequency: 3–30 kilohertz
Wavelength: 100–10 kilometers
Examples: Natural radio signals, communications with submarines and other military communication, emergency beacons, e.g. avalanche beacons.

Name: Ultra low frequency
Abbreviation: ULF
Frequency: 300–3,000 hertz
Wavelength: 1,000–100 kilometers
Examples: Natural radio signals, communication within mines.

Name: Super low frequency
Abbreviation: SLF
Frequency: 30–300 hertz
Wavelength: 10,000–1,000 kilometers
Examples: Communications with submarines, mains/grid AC electrical supply (50–60 hertz), natural radio signals.

Name: Extremely low frequency
Abbreviation: ELF
Frequency: 3–30 hertz
Wavelength: 100,000–10,000 kilometers
Examples: Brain electrical activity, communication with submarines, electromagnetic waves from the Earth's ionosphere.

Name: Sub-Hertz
Abbreviation: sub-Hz
Frequency: <3 hertz
Wavelength: >100,000 kilometers
Examples: Geomagnetic pulsations, electromagnetic waves from the Earth's ionosphere, and possibly earthquakes. Researchers studying earthquakes are looking into the possibility that the detection of sub-hertz and other very low frequency waves could be used as a precursor to an earthquake, as very low frequency waves may be given off by the underground strata as the stresses build up within the strata before the ground moves to produce an earthquake.

The Audio Frequency Range

It is not possible to hear electromagnetic waves. Only sound waves that have been generated by compression waves in the air can be heard. However, whilst discussing different frequencies, and as we shall be dealing with radio receivers, headphones, etc. in later chapters, it seems worth mentioning the audio frequency range.

The audio frequency range, sometimes called AF, is not a true group as it covers a number of frequency ranges. However, it can be useful in its own right, if only for a reference tool.

The audio frequency range is approximately 20 hertz–20 kilohertz. As we grow older our ability to hear the higher frequency range is lost. One's own ability to hear 20 kilohertz has long since gone, and it's only young people who can hear as high as 20 kilohertz. This fact led to an experiment being tried quite recently. The aim of the experiment was to play a loud noise pitched at around 20 kilohertz through a loud speaker in areas where teenagers would congregate in the evenings, such as the outside of shops/stores, etc. The idea was that the constant high pitched sound would only be heard by the teenagers and it would encourage them to more on. But teenagers being the resourceful people they are recorded the sound onto their cell phones and used it as a text message alert, so they could send text messages to each other while in the school classroom, and without the teacher who was older being able to hear a thing. Touché!

The Properties of Different Wavelengths and Frequencies

Radio waves, no matter what the wave length, have the same character, although different laws of propagation. Higher frequencies are more easily reflected from large objects such as hills or large buildings then lower frequencies. Lower frequencies have the ability to penetrate such objects or travel around them. Higher frequency radio waves have a tendency to behave like light waves, because of the shorter wavelength. If they come into contact with an obstruction, for example a large building which is greater in size then the wavelength of the wave, they will be reflected like a light beam hitting a mirror. Lower frequencies, having the longer wavelengths, have the ability to travel around or over such objects. If we take for example a frequency of 3 hertz which has a wavelength of 100,000 kilometers, this wave could easily pass an object the size of the planet Earth without any trouble.

Each frequency range has its own unique qualities. The higher frequencies are useful for sending large amounts of information and as a rule are received by relatively small antennas. The lower frequencies have good penetrating properties and can travel through the deepest oceans for communication with submarines, and have antennas measured in kilometers. Radio astronomers do have a number of protected frequencies which are left clear from traffic, so they can receive the weak signals from space without too much interference. One of these protected frequencies is the 21 centimeters hydrogen line, mentioned earlier.

We have seen that radio telescopes are used for communication with space probes and the sending of messages to any extraterrestrial civilizations that may be listening, but this is just a form of multi-tasking. The same goes for using a radio telescope for sending and receiving radar signals. Radar is a powerful tool in astronomy, and it has been used to map the surface of Venus from Venusian orbit. Radar has been bounced off of other planets in order to obtain accurate rotational periods for a planet, as in the case of Mercury. Radar has even been reflected off the Moon to accurately measure its distance from the Earth.

Using a radio telescope to send out a beam of radio energy and wait for the return echo is radar. To do this energy must be sent out in order for it to return. So energy needs to be created, this energy has been created on the Earth and therefore not of an extraterrestrial origin. When a radio telescope is used to receive radio energy from an extraterrestrial source radio astronomers are interested two types of radio energy or radiation, these are thermal radiation and synchrotron radiation.

Thermal radiation is electromagnetic energy that is given off by a hot object. If we use the spectral classification of stars to illustrate this, blue stars are far hotter that red stars. A "B" class star can range in surface temperature from 10,000 to 30,000 Kelvin, while on the other hand an "M" class star has a surface temperature range of 2,500–3,900 Kelvin. If we set aside the fact that some red stars are huge, as in the case of Betelgeuse, and just stick to the surface temperature, blue stars will emit more thermal energy than their red counter parts. This means that there is a direct correlation between color and temperature. An atom has a certain energy level to start with, this is known as its "ground state". If energy, for example heat, is applied to that atom, or to a molecule (a molecule being a combination of two atoms or more), the electron(s) that are orbiting the nucleus of the atom(s) will become excited, and this will cause them to jump to a higher energy state or orbit.

Likening the solar system to an atom again, with the Sun as the nucleus and the planets as orbiting electrons, the effect of the electron jumping to a higher energy state or orbit would be like the Earth jumping out to the orbit of Mars. This jump to a higher energy state or orbit is just that, a jump and not a slow migration. The energy that is applied to the atom(s) can be steadily increased, but no change will take place until the exact amount of energy needed to make the electron(s) jump to a higher energy state is applied. As the change happens and the electron jumps, it releases a photon at a unique wavelength. Depending on the atom and the amount of energy that is applied to it, as the electron(s) make the jump from one energy level to the next, the photon that is released will always have its own unique wavelength for that particular atom of material. Also the reverse can happen. If an atom loses energy for example it cools down, then the electron(s) jump back to a lower energy state or an orbit, closer to the nucleus of the atom. When this happens another photon will be released, and this too will have its own unique wavelength.

A good example of this phenomenon, and one that is quite easy to do, is to place a piece of copper, such as a soldering iron used to solder sheet metal, (not the type used to solder electrical components) into a gas flame to heat it. When the soldering iron reaches the right temperature for soldering, the gas flame turns a beautiful emerald green color. This is due to the atoms of copper releasing photons at the wavelength that the human eye sees as the color green. This is the same reason why images of the Sun taken in hydrogen alpha light are always a red-orange color, because this is the wavelength of hydrogen alpha.

This is fine for something that can be seen. Taking a piece of metal and with a oxy-acetylene torch, heat up the metal until it starts to glow a dull cherry red, it can be seen that the metal is hot. This is because photons of the color red are being released. If the heating is now stopped, then the metal will start to cool and after a few minutes the dull cherry red color will disappear. The metal will then look visually

2.4 Radio Frequencies and Characteristics

the same as it did before it was heated, but the metal will still be hot enough to burn one's hand if it is touched. One will still be able to feel the heat radiating from the metal. This is infrared radiation, but because our eyes are not sensitive to infrared radiation it cannot be seen. In space the average temperature is 2 or 3 degrees above absolute zero. If this experiment was conducted again, but in space, even if the metal has cooled below the temperature that can be detected within the infrared part of the spectrum, the metal will still be radiating energy. This is as long has its temperature is above the average temperature of space. The metal will radiate energy, but now only in the longer wavelengths of the radio part of the electromagnetic spectrum.

Whilst studying radio astronomy the term "black body" or "black body radiation" may crop up, especially in books that deal with radio astronomy and spectroscopy. This may seem a little complicated, but simply put: a black body is a theoretical object that absorbs thermal energy (heat) and is a perfect emitter of thermal radiation. The characteristic of thermal radiation is that it has a continuous spectrum.

Thinking of the above example of heating the piece of metal. The metal can be thought of as a black body adsorbing thermal radiation as it is heated, and the energy that the metal releases when no longer being heated as being black body radiation.

This can be seen by the hissing noise that Robert Wilson and Arno Penzias received from their horn antenna. This was the thermal radiation left over from the big bang at the start of the universe which has now cooled from the unimaginable temperature at the moment of the big bang to 2 or 3 degrees above absolute zero over the last 13.75 billion years.

A good example of a source of thermal radiation that we can pick up using a radio receiver such as the Radio Jove receiver would be thermal radiation from the Sun.

The next type of electromagnet radiation that radio astronomers are interested in receiving is synchrotron radiation. So named because it was first observed in particle accelerators called "Synchrotrons". The characteristics of synchrotron radiation compared to that of thermal radiation are quite different, and this makes synchrotron radiation easier to detect. This type of radiation is produced by particles, usually electrons, accelerated to high speeds within a magnetic field. This may sound familiar, as the search for the Higgs Boson at Cern using the Large Hadron Collider (LHC) has been using high-powered magnetic fields to accelerate particles up to 99 % of the speed of light and smashing them into each other and imaging the particles that have been created by these collisions. The planet Jupiter, with its vast and powerful magnetic field, would make a rather good particle accelerator, but it may prove a little bit difficult to control.

If we take the example of a magnetic field and an electron spiraling along a magnetic field line within the magnetic field. As the electron spirals along the field line it produces electromagnetic radiation. The frequency of the emission is directly related to how fast the electron spirals, the faster the spiral the higher the frequency. If a stronger magnetic field is used, this has the effect of tightening the spiral for the electron and therefore increasing the frequency of the electromagnetic radiation produced. Furthermore, due to this spiraling motion synchrotron radiation emissions are polarized.

In radio astronomy synchrotron radiation comes from some of the most powerful sources in the universe, such as radio galaxies. M87 in the constellation of Virgo is a good example of a radio galaxy. This radiation has also been detected from quasars and supernova remnants, such as the Crab Nebula, in particular the high energy particles coming from the central pulsar. Also particles that get trapped within Jupiter's powerful magnetic field.

Chapter 3

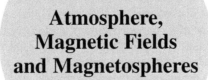

Atmosphere, Magnetic Fields and Magnetospheres

3.1 The Goldilocks Zone

A planet's atmosphere is a key part of its make up, but sometimes we can overlook it or take it for granted, especially on the Earth. If we were to look at the four terrestrial planets Mercury, Venus, Earth, and Mars, we would see a vast difference in the atmosphere of each.

It was first thought that Mercury didn't even have an atmosphere, but thanks to the MESSENGER (MErcury Surface, Space ENvironment, GEochemistry, and Ranging) space probe, Mercury has been discovered to have a very wispy atmosphere, but nothing of any structure and substance. The planet still gets baked on the sunlit side, and frozen on the opposite side. The Sun's radiation has irradiated Mercury's surface with lethal doses of radiation and charged particles.

The planet Venus looks so bright in the sky, it is understandable that ancient civilizations thought it beautiful and named it after the goddess of beauty. But with its dense atmosphere of almost pure carbon dioxide, crushing atmospheric pressure 90 times that of the Earth, the heat at the surface is hot enough to melt lead, not to mention the sulfuric acid in the clouds. It's more like a vision of hell than of beauty.

Meanwhile, the atmosphere on planet Mars is so thin it barely exists, and so cold that carbon dioxide freezes out of the atmosphere in its winter and collects at the poles. Mars has global dust storms, which can last for weeks, and seem to always happen when the planet is at its closest apparition to Earth, meaning nothing on its surface can be observed with an optical telescope.

Whoever coined the phrase "the Goldilocks Zone" about the Earth being not too hot and not too cold but just right was correct. The Earth's atmosphere is more than just a mixture of approximately 78 % nitrogen, 21 % oxygen and the last 1 % a mixture of carbon dioxide and other gases. Without the atmosphere there would be no life on Earth. It has the ability to protect us from the vacuum and extreme cold of space, and the unimaginable heat from the Sun. Our atmosphere also stops the dangerous electromagnetic wavelengths from space getting through. As a percentage the atmosphere doesn't add up to much, but we would be in trouble without it.

Earth's atmosphere is made up of a number of layers, with transition layers between each. The five major layers are as follows: (Please note: All altitudes used here are approximate as they can vary from season to season and with the latitude of the observer on the Earth). The first layer is the troposphere, It starts at sea level and rises to 12 kilometers (7.5 miles) in altitude. This is the thickest part of the atmosphere and where the air pressure is at its greatest, and also where most of the weather occurs. This is what we call home.

The stratosphere is the second and next major layer and covers the range from 12 kilometers (7.5 miles) to 50 kilometers (30 miles). This layer contents the ozone layer at 20 kilometers (12.5 miles) to 40 kilometers (24.8 miles). The mesosphere the third major layer this starts at 50 kilometers (31 miles) and ends at 85 kilometers (52.8 miles). This is the altitude at which the rare noctilucent clouds are seen and were meteors burn up. Also the start of the layer, the ionosphere, that will affect our radio astronomy projects. The forth major layer is the thermosphere. This layer starts at 85 kilometers (52.8 miles) but due to solar activity it fluctuates in altitude between approximately 350 kilometers (217 miles) and 750 kilometers (466 miles). This is also the start of the heterosphere. The heterosphere is where the gases in the atmosphere start to separate into component parts and the mixture that we call "air" is lost. The lighter trace gases such as helium and hydrogen can simply float away. The last major layer is the exosphere. This is the outer most layer of the atmosphere and the realm of the two gases hydrogen and helium, but the gases at this altitude are so rarefied that they barely interact with each other.

The atmosphere plays a large part in blocking certain wavelengths from getting through to the Earth's surface. As mentioned in the last chapter the shorter wavelengths carry a lot of energy and can be very harmful. The Fig. 3.1 shows the effectiveness of the atmosphere in stopping these harmful wavelengths.

Earth's atmosphere blocks everything from gamma rays right up to the longer wavelengths of ultraviolet just before the visible wavelengths. It is then transparent to visible light but it starts to become opaque in the infrared with only a few wavelengths getting through. The long wavelengths of the far infrared are blocked, and then it becomes transparent for parts of the radio wavelengths. If we look at the altitude at which each wavelength is blocked it can be seen why satellites and space telescopes are needed to effectively observe wavelengths such as ultraviolet, and X-ray. But some useful astronomy can be done with high altitude balloons for the X-ray wavelength. Some useful infrared astronomy can be done from the ground, but water vapor in the atmosphere blocks a lot of infrared wavelengths getting through. One way around this is to use airborne telescopes. One such telescope is NASA's (SOFIA) (Stratosphere Observatory For Infrared Astronomy). This airborne telescope can fly at an altitude where it is above 99 % of the water vapor in the atmosphere, and has proven to be very useful at observing at infrared wavelengths.

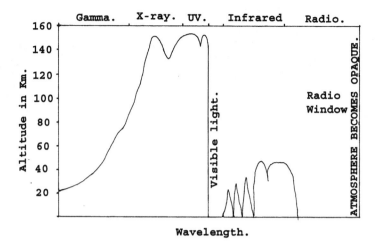

Fig. 3.1 This image shows the effectiveness of the atmosphere at blocking short wavelengths of electromagnetic radiation

3.2 The Ionosphere

Radio waves were thought to travel in straight lines or be a line of sight phenomena, but when Marconi did his trans-Atlantic transmission in 1900 it was realized that some sort of reflection must be taking place to get the transmission around the curvature of the Earth. In the mid 1920s experiments were carried out which proved the existence of the ionosphere, and a mathematical model was put forward. But the big leap came with the first artificial satellite Sputnik 1, which was launched in 1957. This satellite, broadcasting at 20 and 40 megahertz and its "beep, beep" signal, could be heard when the satellite orbited over head. This proved without question that the ionosphere could be penetrated if the right frequency was used. The launch of Sputnik 2 with Laika on board further narrowed down the frequencies that scientists identified were able to travel through the ionosphere.

The ionosphere is very important for three of the projects covered in this book, and we will use the properties of the ionosphere in different ways. A vast amount of work has gone into the study of the ionosphere and a large amount has been written on the subject. Radio enthusiasts around the world use the reflective qualities of the ionosphere to bounce their signals great distances to other radio enthusiasts, in a practice known as DX communication or "DXing". But to use the ionosphere, we need to know what it is and how it works.

The ionosphere is a shell that surrounds the Earth. It starts at an altitude of approximately 70 kilometers (43.5 miles) and extends to more than 600 kilometers (373 miles). It owes its existence and variability to the high energy wavelengths from the Sun, such as ultraviolet and X-ray. It can change from season to season, also as the result of solar activity. For example, the Sun's 11 year solar cycle from

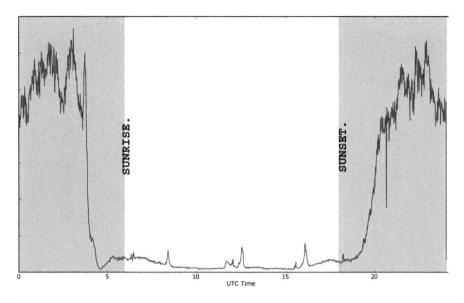

Fig. 3.2 A day of SuperSID data used to show changes to the ionosphere

solar maximum to solar minimum, in particular such activity as solar flares can produce a marked effect on the ionosphere. There are seasonal variations to the ionosphere for an observer on the Earth. For example, an observer at 50 degrees latitude will notice a change from midsummer to midwinter as the height of the Sun changes in altitude from month to month.

The ionosphere can even change over a few hours from day to night. A good way to demonstrate just how quickly it can change is to look at the Fig. 3.2. The image is of a day's worth of SuperSID data (the SuperSID monitor will be discussed later). There is no need at this stage to understand the graph, as we are only interested in the change to the ionosphere shown by the graph. At night there is very little ionization. This can be seen by the height of the graph in the night time hours. Looking at the sunrise and sunset sides which have been marked on the graph, with time across the bottom axis, it can be seen that the ionization of the ionosphere at sunrise happens almost instantly, as demonstrated by the near vertical line on the sunrise side of the graph. On the sunset side of the graph, the line is at an angle, indicating that it takes a little longer (a couple of hours or so) to change and lose the ionization (Fig. 3.2).

The ionosphere is made up of a number of layers. The only layers that are of interest to us are the "D" layer, "E" layer, and "F" layer. Each layer is ionized by different types of radiation. The inner "D" layer is affected by hard X-rays, the "E" layer with soft X-rays and extreme ultraviolet, while the outer "F" layer is ionized with extreme ultraviolet and near ultraviolet. During the day the side of the Earth facing the Sun is fully exposed to the radiation emitted by it. This causes a large increase in the ionization of

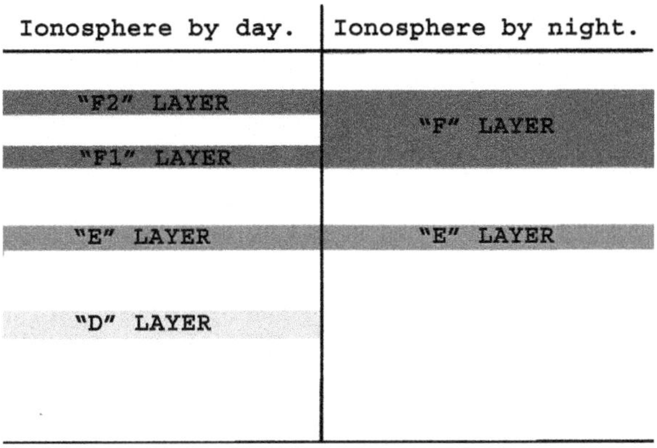

3.2 The Ionosphere

Ionosphere by day.	Ionosphere by night.
"F2" LAYER	"F" LAYER
"F1" LAYER	
"E" LAYER	"E" LAYER
"D" LAYER	

Fig. 3.3 The change in the layers of the ionosphere by day and night

the atoms in the "D", "E" and "F" layers. The outermost "F" layer separates into two layers, F1 and F2. See Fig. 3.3.

This separation of the F1 and F2 layers during the daytime allows for long distance communications at high frequencies, as radio waves can be reflected greater distances by the higher F2 layer. The E layer is accessible to wavelengths of 80 to 2 meters, and the D layer in the daytime absorbs wavelengths of between 160 and 40 meters. As can be seen the effects of this ionization has a lot to do with the wavelengths that are being used, either the shorter or longer wavelengths. As different wavelengths are effected in different ways by this ionization. For our purposes this ionization is excellent news for the SuperSID monitor as it relies on the reflective qualities of the ionosphere to function.

For the Radio Jove receiver this ionization makes it almost impossible to pick up Jupiter during the day, although some people have managed it if Jupiter is high in the sky and the Sun's activity is relatively low, or the Sun is low in the sky. For example midwinter so the ionization within the ionosphere may be lower than normal. But as a rule the signals from the planet Jupiter hit the ionization and are reflected back into space and are lost. This ionization is less of a problem for the radio emissions from the Sun as they are strong enough to force their way through the ionization, and this allows us to receive them. The ionization affects on the INSPIRE receiver trap the very low frequency signals between the Earth and the ionosphere and bounce them between the two, stopping the signals travelling higher than this ionized layer and limiting the types of sounds that can be heard.

As mentioned above the level of the ionization within the ionosphere can soon change. A couple of hours or so after sunset the ionosphere can "cool down" and the two "F" layers recombine as they lose some of their ionization. The "E" layer also loses most of its ionization, and the "D" layer can disappear completely. In the case of the SuperSID monitor, when the Sunsets and the ionosphere loses its ionization, the monitor ceases to function because it relies on the reflective quality of the ionosphere to work. But the Radio Jove receiver at this point comes into its own as it is now able to pick up the radio emission from the planet Jupiter, as these radio emissions can now pass through the ionosphere to our receivers. Although if solar activity is high, maybe there are a large number of sunspots on the Sun. The ionosphere can remain partly ionized all night, and this makes receiving Jupiter's radio emissions very difficult if not impossible. With the ionization reduced, the very low frequency signals received by the INSPIRE receiver can travel out into space and get trapped in the Earth's magnetic field, and travel great distances from one hemisphere to the next. This then produces the haunting sounds of whistlers.

To summarize, the ionosphere can be our friend or our foe, it all depends on the type of observing that we wish to do. The SuperSID monitor needs the ionization within the ionosphere to function correctly and when the ionization is lost it ceases to work. On the other hand, the Radio Jove receiver can be affected quite badly at night if the ionization is still quite heavy. If a large number of stations can be heard, or if struggling to find a quiet frequency to listen for Jupiter, this is a sign that the ionization is still quite bad. The INSPIRE receiver will still receive signals but just of different types depending on the state of the ionosphere. These effects will be covered in greater detail in the relevant chapters for each project.

3.3 Scintillation

Another effect of the Earth's atmosphere is scintillation. As any optical astronomer knows, if an object, especially a bright one is observed close to the horizon the object can twinkle. A good example of this is the star Sirius. Sirius, when viewed through binoculars, can seem to come in and out of focus and can also change its color quickly. This is because it is being observed through the thickest part of the atmosphere. Also there are partials of dust and pollution to contend with as well. All of these factors play a part in the "seeing" of an object. As a general rule all objects within 15–20 degrees of the local horizon are prone to this unless the "seeing" is very good. The higher an object is in relation to the observer's horizon the better.

This is also true of radio astronomy, as the radio waves travel towards the Earth's atmosphere they are travelling parallel to each other. But as they travel through the ionosphere, the ionization can be irregular; there may be more in one area and less in others. The radio waves leaving the ionosphere can be affected by the irregular ionization, in the same way that the light from a star seems to make it twinkle. Also the radio emissions from an object, for example the planet Jupiter, are less troubled if the height of the planet in relation to the observer is greater, as the antenna is less likely to pick up interference and other stray signals if the center of the antenna beam is set at 50 degrees than if it is set for 20 degrees above the local horizon.

A good example of scintillation is to look at the lines at the bottom of a swimming pool marking out each of the lanes. Looking at one of these lines through the water the line will appear to distort its shape, become blurred and then, refocus and in some cases disappear briefly as the water reflects the light straight back to the observer. The water is acting in the same sort of way as the Earth's atmosphere and distorting any electromagnet radiation passing through it.

3.4 Planetary Magnetic Fields and Magnetospheres

A planet's magnetic field can also play an important role in our radio observations, especially those of the Earth and Jupiter. If we now look at the four terrestrial planets we shall see how the Earth is quite unique in having a strong magnetic field. But even the strong magnetic field of the Earth is dwarfed by the gas giant Jupiter.

Out of the four terrestrial planets Mercury, Venus, Earth and Mars, the Earth has the greatest magnetic field. Mercury has a small magnetic field despite its size, but its magnetic field is only around 0.01 times that of the Earth's. The magnetic field is thought to come from a large nickel iron core, but this core is now thought to be solid as Mercury is too small for it to have hung on to its internal heat. This magnetic field, although weak, is able to stop some of the charged partials from the solar wind, but the lack of any substantial atmosphere means that other particles can get through to the planet's surface and bathes Mercury in lethal doses of radiation.

Venus is a real mystery. A planet roughly the same size as Earth, Venus should have a magnetic field similar to that of the Earth, but in reality it has very little or none. Space probes haven't managed to measure a magnetic field as yet because it is that weak. The best estimates are that Venus does have a magnetic field but it is in the order of 0.00001 times that of the Earth. Venus is large enough to have kept its internal heat, and in fact is still thought to have volcanic activity, but because of Venus's very slow rotation (243 Earth days) it is thought that there isn't the dynamo effect as on the Earth. Without a protective magnetic field, Venus's atmosphere is slowly being lost to space due to the action of the solar wind. Space probes that have ventured inward towards Mercury and Venus have observed a stream of gases coming from Venus in a similar way to the stream of ices and gases that produce the tail of comets, as Venus is slowly losing its atmosphere.

Mars has a magnetic field only 0.002 times that of the Earth. Mars rotates fast enough (just over 24 hours) to have a dynamo effect, but because of its small size this has probably led to Mars losing its internal heat resulting in the solidification of the core.

As we can see, the Earth's magnetic field is a lot more powerful than the other three terrestrial planets. The magnetic field of the Earth can be likened to that of a bar magnet, with field lines similar to the ones discussed in an earlier chapter about magnetism around a coiled wire. The magnetic north and south poles are offset to the rotation of the Earth's axis by approximately 12 degrees. This can change annually very slightly, and large earthquakes have been known to make slight changes too. The Earth's magnetic field has flipped in the past, in a similar way in which the Sun flips it magnetic field every 11 years or so. No one really knows why this happens for sure. Samples taken from very old lava flows have been examined, and it has been found that small pieces of magnetic rock within the lava have lined themselves up with the Earth's magnetic field lines at the time the lava solidified, and this alignment is of opposite polarity to the magnetic field of the Earth today. It would be interesting to know how birds would cope with the reversing of the magnetic field. Migratory birds have magnetic cells within their brains that line up with the magnetic field lines of the Earth, and this allows them to navigate the thousands of kilometers between their breeding grounds.

The origin of the magnetic field lies within the Earth's core. The inner core, thought to be iron, is solid, and this floats inside the outer liquid core, thought to be a mixture of nickel and iron. The liquid outer core has the consistence of thick treacle. Convection currents within the liquid outer core are generated as the hot material rises and drops back down in a circular motion as the material cools. A good way to imagine this is to think of those lava lamps that were very popular some years ago. The heat from the electric light bulb heats the wax, and this then causes the wax to rise through the oil in the lamp. As it cools, the wax sinks back down again and the process repeats. The inner core rotating within the liquid outer core causes these convection currents, and combined with the speed of the rotation of the Earth acts like a large dynamo which produces the vast magnetic field of the Earth. This magnetic field gives us the magnetosphere.

The Earth's magnetosphere extends many thousands of kilometers into space and its distance varies greatly mainly due to its interaction with the solar wind. The magnetosphere also contains within it two large torus-shapes areas, these are like two huge tire shaped zones that surround the Earth. There is an inner and outer torus. These are the Van Allen radiation belts, named after James A Van Allen who first theorized their existence. Please see Fig. 3.4.

These radiation belts are made up of a highly charged plasma that has been energized by charge particles from cosmic rays, but mainly from the charged particles contained within the solar wind. These particles then become trapped within the Earth's magnetic field. The outer belt contains mainly electrons from the solar wind, the inner belt contains the higher energy particles from the solar wind, but more importantly it also stops the highly charged particles that come from cosmic rays and prevents them reaching the Earth's surface.

The magnetosphere acts like a large buffer against the solar wind. Charged partials from the Sun hit the outer belt and are channeled along the field lines away from the Earth. In times of high solar activity the incoming pressure of the solar wind on the Sun side can increase greatly and cause more deforming of the outer belt. The side away from the Sun can develop a long magnetic tail that can stretch many thousands of kilometers into space. As more and more energy from the solar wind hits the leading edge of the magnetosphere the tension builds up in the long magnetic tail like an elastic band, and at some point the elastic band snaps back. This brings charged particles careering along the field lines and funnels them down at the poles where they interact with the atoms in the atmosphere. This produces the aurora. If there are large amounts of aurora activity this can heat the ionosphere and cause it to expand into the magnetosphere. Therefore, if a large amount of energy were to hit the magnetosphere, for example an X-ray flare from the Sun, the

3.4 Planetary Magnetic Fields and Magnetospheres

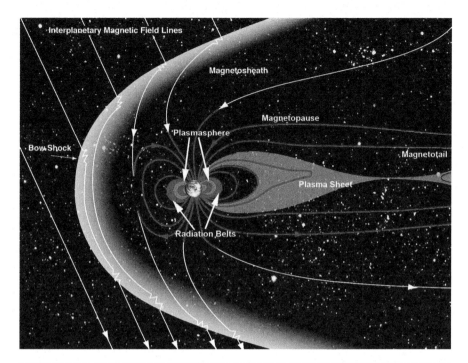

Fig. 3.4 The Earth's magnetosphere

magnetosphere would distort and in turn this would distort the ionosphere. This is known as a sudden ionospheric disturbance or SID. As mentioned earlier, if a low frequency wave gets through the ionosphere it can get trapped within the magnetic field lines, and it can travel hundreds of miles into space before being brought back down the magnetic field line. This can allow them to travel great distances even from one hemisphere to the next. These are known as whistlers.

Jupiter's Magnetic Field and Magnetosphere

Since the planet Jupiter is on our list of radio astronomy targets for observation with the Radio Jove receiver, it would be a good idea to look at the planet through the eyes of the radio astronomer as well as the eyes of the visual astronomer.

When looking at Jupiter visually through an optical telescope it is a wonderful object to observe and study. Its sheer size within the eyepiece of a telescope makes it an easy object to observe. Even with the smallest aperture telescope it can be seen that the planet is not circular but squashed in at the poles. This is due to its fast rotation, a little under 10 hours. In larger aperture telescopes the detail visible within the cloud belts themselves is simply breath taking, and if the quick rotation is taken into consideration the details within the cloud belts soon change. Jupiter is a dynamic planet with an ever-changing face. No wonder then that Jupiter is a favorite target for planetary-imagers. Many have engaged in hours of conversation speculating about the recent disappearance and reappearance of one of the many cloud belts, discussing what caused it to fade and then come back again. The planet's most famous feature, the great red spot, is a huge storm even bigger than the planet Earth. First observed over 300 years ago, it is still raging around the planet today and has been observed devouring smaller storms which have the misfortune to get in its way. This storm has a peculiar trick that it sometimes performs of changing its color. How and why this color change takes place no one really knows, but it has been observed to fade to a pale pink and then reappear red again. Some theories suggest that there is a hole in the upper atmosphere which acts like the eye of a huge hurricane which produces a tall column leading deep down into the vast atmosphere. Other ideas are that the spot floats up and down within the atmosphere, as the density of the atmosphere changes with heat. This can be likened to the action of a lava lamp, as the wax is heated it floats to the top and sinks back down again as it cools. This could account for the change in color of the red spot: as the spot sinks lower it loses some of its color and becomes paler, but as it climbs through the atmosphere again it regains its familiar red color.

While finalizing the Voyager 1 and 2 space probe's flight plans in the 1970s, it was just by sheer chance that one astronomer suggested paying a visit to the four largest Jovian (Galilean) moons to observe them as they perform their endless dance around Jupiter. The four Galilean moons are Io, Europa, Ganymede and Callisto. These four moons were first thought to be like our own Moon, dry and arid and not worth any mission time to observe them. How wrong they were. The innermost moon, Io, is the most volcanic body in the solar system. Io is subjected to massive tidal forces from Jupiter and the other Galilean moons, producing vast flexing of the whole moon. This flexing heats the interior of Io and drives its volcanism. Io was the first body where volcanic eruptions were observed other than the Earth. Any map of Io would be out of date in a matter of a few weeks, due to the surface being re-laid over and over again. Its multicolored surface of sulfurous deposits has earned Io the nickname of the flying pizza.

When the Voyager space probes flew past Europa, a moon as smooth as a billiard ball, it detected evidence that there may be a salt water ocean below the ice covered exterior. That ocean could be kept as liquid water rather than ice by the tidal heating forces that Europa receives from Io and Ganymede, with the ice acting as a barrier to the high doses of radiation from Jupiter. Could this be a good place for life to start? Plans are on the table for a mission to Europa hopefully in the not too distant future. That mission will seek to land a craft on the ice and release a probe to melt its way through the ice uncoiling a long antenna as it goes. This probe would then release a small submersible to search for life in the salt water ocean. What a triumph it would be if life is found below Europa's icy crust! Ganymede, the largest moon of Jupiter and the largest moon in the solar system, is bigger than the planet Mercury. Its surface is covered with areas of dark and light material thought to be a mixture of mainly ice and carbonaceous compounds. It has bright rays extending

3.4 Planetary Magnetic Fields and Magnetospheres

from its cratered surface. Callisto, the outermost of the four Galilean moons, is also larger that the planet Mercury. It has a heavily cratered surface and a large multi-ringed basin called Valhalla.

The descriptions of Jupiter and its moons outlined above are all in visible light wavelengths, and therefore only half the picture can be seen. Looking at the planet Jupiter in the radio wavelengths opens up a whole new field of study. Remember, from the above image, how far the Earth's magnetosphere extends into space. Now consider Jupiter and its distance from the Earth. Jupiter observed using unaided eyes is seen in the sky as a bright off white colored −2 magnitude star like object. Some very keen sighted people have even managed to pick out a couple of the four Galilean moons. If it was possible to view Jupiter's magnetosphere using only unaided eyes, it would be seen to cover an area greater in size than that of the full Moon in the sky. Jupiter's magnetosphere is in fact the largest planetary magnetosphere in the solar system, even bigger than the Sun itself.

When solar activity is high, and the solar wind buffets Jupiter's magnetosphere on the side facing the Sun, this can produce a magnetic tail on the night side that can stretch right out to the orbit of the planet Saturn. A quick search on the internet, at the NASA solar system exploration website will find a number of sound samples recorded by the two Voyager spacecrafts as they flew past Jupiter and Saturn. Two of the sound samples are Voyager 1 as it negotiates Jupiter's powerful magnetosphere and Voyager 2 picking up lightning discharges in Jupiter's atmosphere, to list but a couple. The intensity of the magnetic field measured from the cloud tops of Jupiter is in the region of 14 times that of the Earth. This may not sound a great difference to that of the Earth but once the size of Jupiter is taken into consideration this makes a huge difference. The secret to this phenomenal magnetic field and magnetosphere is contained within the internal structure and composition of Jupiter itself.

Jupiter is sometimes referred to as a failed star. Theories have suggested that if Jupiter was approximately ten times larger, the gravitational forces developed by its mass would be great enough at the core that it would be enough to start nuclear fusion within the core itself. Not at the rate that happens at the center of the Sun, but as a brown dwarf. There is some conjecture whether this would be the case or not, but Jupiter at the size it is now radiates twice as much energy as it receives from the Sun. There has been many theories why this is the case, one of which is that Jupiter's solid core is slowly shrinking. Looking at the structure of Jupiter starting from the outside and working towards the core, it will be seen how Jupiter develops its powerful and extensive magnetic field. The planet's atmosphere is mainly a mixture of mostly hydrogen and helium, with a small percentage of other elements. The atmosphere extends many thousands of kilometers down to its solid core. Travelling through the atmosphere, the pressure increases and we reach a transition point where hydrogen ceases to be a gas and starts to change into a liquid. Traveling further down, all the hydrogen gas has gone and there is just liquid hydrogen. Travel still further and there is another transition zone where the atmospheric pressure is now so great that the hydrogen liquid starts to turn into a more exotic liquid, metallic hydrogen. Even further down one reaches the point where all the liquid hydrogen has turned into liquid metallic hydrogen. The last stop on the journey is what is thought to be a rocky core.

Remembering the discussion of the Earth's magnetic field, it is hypothesized to result from hot molten metal in the outer core of the Earth. Jupiter owes its own magnetic field to the large amount of liquid metallic hydrogen in its outer core. Recall that the Earth has a dynamo effect which is facilitated by convection currents within the outer molten core and the effect of the Earth's rotation. Jupiter has the same convection currents in the liquid metallic hydrogen outer core as the Earth does in its outer core. Coupled with the faster rotation period of Jupiter (under 10 hours, or more than twice that of the Earth), not to mention the sheer size of Jupiter compared with the Earth, this makes for a very powerful dynamo. It produces a magnetic field 20,000 times more powerful than that of the Earth, resulting in Jupiter's vast magnetic field and magnetosphere. Jupiter's innermost moon, Io, orbits very close to Jupiter and well within the magnetic field. It has been suggested that the action of Io orbiting at this distance causes an electric current to be produced somewhere in the order of a billion amps. Standing on the surface of Io this electric current would make for a really good show of aurora in Io's wispy sulfurous atmosphere, but the show couldn't be enjoyed for long before receiving a lethal dose of radiation from Jupiter!

Radio emissions can come from Jupiter itself. There are three regions that are of interest to anyone with a Radio Jove receiver, known as the A, B, and C regions. These areas revolve with the planet and are not always in the right place to be picked up from Earth, as will be covered in greater detail within the Radio Jove chapter. Radio emission can also be produced by particles from the solar wind getting trapped within the magnetosphere and spiraling down the magnetic field lines while releasing energy in the form of radio waves so powerful that they can be picked up from the Earth using the Radio Jove receiver. Io also has a part to play in these radio emissions, in a phenomenon called the Io effect. As Io orbits Jupiter, all the electrical energy produced has the effect of concentrating any radio emissions into a large cone shape. The cone travels ahead of Io in its orbit, and has the result of concentrating any radio emissions into a sweeping beam, similar to but not as narrow as the beam from a lighthouse. If the Earth is in the path of the beam, any radio emissions should be of greater intensity than if Io was at another point in its orbit.

Chapter 4

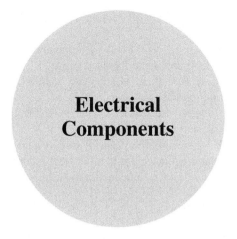
Electrical Components

4.1 A Brief Review of Electrical Safety

ELECTRICITY CAN KILL! If in any doubt about mains/grid power supply consult a qualified electrician.

It is impossible to think of every eventuality, but if a few common sense precautions are put into practice then no harm should come from building and using these projects.

Remember it can get damp outside in the evening and early morning with dew from the grass and condensation as things cool down. Therefore it isn't advisable to have mains/grid power outside on an extension lead, as there is a risk of electric shock. There is the added danger that someone may trip over the lead in the dark.

If a soldering iron is being used make sure it's not within the reach of children or pets, and give it at least 30 minutes to cool down after use before putting it away.

If main/grid power is to be used instead of batteries, seek the advice of a qualified electrician if in any doubt.

Please follow any instructions supplied with the kits carefully, especially where polarity is involved. Some electronic components are polarity sensitive. As a rule of thumb: check twice, solder once.

Replace any blown fuse with one of equal value.

If there is a local thunder storm, switch off any receivers/monitors, computers, and unplug antennas.

Do It Yourself (D.I.Y.) Safety

Take the precaution of wearing eye protection and other safety protection as and when the need arises.

If power tools are to be used, refer to the manufacturer's instructions and follow any advice given. The same applies to hand tools, etc.

When soldering, make sure there is adequate ventilation in order to remove fumes from the soldering process.

Solder now comes in lead-free varieties, but it is still good practice to wash hands after use or before eating or drinking.

When working at heights, e.g. off a mast, be sure to wear the correct fitting harness. If working off a ladder, make sure it is the right ladder for the height. Don't over reach, and do be sure to have someone trustworthy steadying the ladder.

Using Equipment in the Countryside

If equipment is to be used in the countryside please respect 'no entry' and 'private land' signs. Get the land owner's permission first before crossing their land.

Each country has its own laws regarding personal safety and the level to which this can be upheld. Take simple precautions such as carrying a cell phone and letting someone trustworthy know the plan and an estimated time of return. But remember to stick to this plan.

It's also a good idea to carry extra food and water, plus a basic first aid kit, and possibly also insect repellent, etc.

In the event of a thunder storm don't be tempted to shelter under a tree. Trees are basically full of water and will make a good lightning conductor. The safest place is inside a metal bodied vehicle.

4.2 Soldering

Soldering Irons

There are a bewildering number of soldering irons available in the market place today. Here are a few pointers to help make a good purchase.

Try to avoid the cheaper unregulated ones, as these can overheat and soon burn out. Also avoid the ones that run off of a 12 volt car battery. Choose a thermostatically controlled one, as these automatically switch on and off after reaching the correct temperature.

A soldering iron's power is measured in Watts, a good choice would be one around 25–35 watts. This would be adequate for all the projects in this book, with

4.2 Soldering

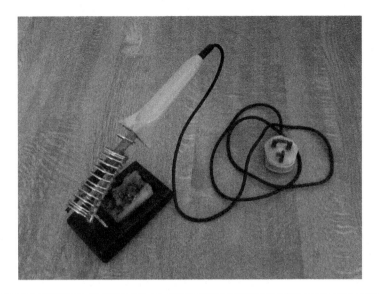

Fig. 4.1 A well used soldering iron with its stand

the exception of the construction of the Radio Jove antennas which need an iron of around 75 watts to heat the heavier gauge wire that is used.

Most good soldering irons come with the ability to change the tip. The tips come in many different sizes and shapes. A good choose of tip would be one of around 2.5–3 millimeters with an angled end rather than a point. Another important accessory is a stand, these stands have a little sponge in the base which is kept damp. Before each new soldering joint give the irons tip a quick wipe (don't waste money buying the replacement sponges just cut a piece off a dish washing sponge – it works just as well). This removes any scum off the tip before making a new joint (Fig. 4.1).

Tinning a soldering iron, it is a good idea at the start of each soldering session to allow the iron to reach the correct temperature, then give the tip a quick wipe on the damp sponge and apply a small amount of solder to the tip. The flux in the solder cleans and freshens up the tip before use, and this needs to be done only once per use.

Types of Solder

There are a number of different types of solder. Always choose one with a non-corrosive flux, as the acid flux will corrode the joints over time.

Solder used to contain large amounts of lead, but the modern day solders are lead free, and come in a number of different diameters. The 0.8 and 1 millimeters are useful sizes to use.

How to Solder

Good soldering skills are essential for electrical work. A poorly soldered joint can make the difference between a project that works and one that does not. The first thing to remember is to keep everything clean and free from any grease, oil, or melted insulation which will stop the solder from doing its job. Any dirt and grease can be removed with a solvent cleaner or a light sanding from a piece of very fine sandpaper. In some books steel wool is suggested, but if this is used and any fibers of the steel wool go adrift there is the potential for a short circuit.

The next important thing to remember is: never melt the solder with the soldering iron. Place the soldering iron on the joint for a few seconds and let the heat from the joint melt the solder. Keep the soldering iron there for a few seconds and allow the solder to fill the joint, then remove the soldering iron and allow the joint to cool. A lot can be learnt from the appearance of a soldered joint. It should look smooth and shiny, almost wet looking. If it looks like this the chances are it's a good joint, but if it looks dull or full of holes it's probably a bad joint. This can be caused by being too eager and not letting the soldering iron reach the correct temperature before using it, or some sort of contamination within the joint or on the soldering iron, or even a bad batch of solder.

The best way to learn how to solder is to have a go. A good place to start would be to get some "stripboard" from a local electronics supplier, as this is ideal to practice the art of soldering. Stripboard has pre-drilled holes in it ready for components to be soldered in place. It is insulated on one side and has lines of copper tracks running its full length on the other. It is used as a base for building amateur electronic projects. Also needed will be some solid core wire of a diameter of about 0.6 millimeters. Strip the insulation off the wire to about 300 millimeters (12 inches) and cut the bare wire into pieces of about 30 millimeters (1.25 inches) bend the last 6 millimeters (0.25 inch) over so that it looks like a big staple.

Place these pieces through the holes on the stripboard so that the ends of the wire appear on the side where the copper tracks are. Now try and solder this into place without getting solder on the other tracks from where the wire sticks through. Try this with different lengths of wire.

If solder is put in the wrong place there are two ways to remove it. One is the use of a de-soldering pump; this is like a spring loaded syringe in reverse. The plunger is pushed down and a clip holds it, the soldering iron is placed on the joint, and when the solder becomes molten the de-soldering pump is placed at the side of the joint and the button holding the plunger down is pressed. This allows the plunger to return to its original position with a slurping action that sucks the solder away from the joint. The second option is to use de-soldering braid. The braid is like a flattened stranded wire approximately 3 millimeters (0.125 inches) wide, and is impregnated with flux. The braid is placed over the area where the solder is to be removed and the soldering iron is placed on the braid from above. The braid soaks up the solder like a sponge. This braid should be removed whilst still hot to prevent the braid sticking to the circuit board.

4.2 Soldering

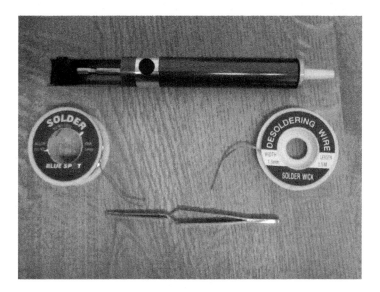

Fig. 4.2 Soldering equipment: *Top center* de-soldering pump, *right* de-soldering wick, *center bottom* "heat shunt" tweezers, *left* 1 millimeter diameter solder

Holding Components in Place

The holding of components in place while soldering has always posed a problem. One way round this problem is to purchase a "heat shunt". Despite its fancy name, a heat shunt is little more than a pair of tweezers that work the opposite way to normal tweezers. Squeeze a normal pair of tweezers and they close, squeeze a heat shunt and they will open. These are great for holding components in place while soldering. They have another equally important use, some components such as transistors are sensitive to heat so heat shunt tweezers are clipped onto the part of the transistor that is being soldered and the heat is shunted away. Small metal crocodile clips and bull clips can also be used to hold components in place (Fig. 4.2).

Holding the Circuit Board While Soldering

Another problem is the holding of the circuit board while soldering. When soldering it may be felt that one pair of hands is not enough. Soldering iron in one hand, solder in the other, then there's the components to hold and on top of that there's the circuit board as well. Four jobs only two hands. So we need a "helping hand". The image below is of a helping hand – this tool costs very little, and the magnifying glass is just an added bonus (Fig. 4.3).

Fig. 4.3 A helping hand tool holding a piece of stripboard

4.3 Multimeters

Although not essential for the projects covered in this book, a multimeter can make life a whole lot easier when dealing with electronics. Just by their shear ability to find faults and the range of tests that they can perform. Multimeters come in two types, analog and digital.

Analog Multimeters

These are usually cheaper to purchase, but it can take a little time to understand how to make a reading. The needle moves across the dial and there are a number of scales to read from. The trick is to remember to read from the right scale that matches the test that's being performed.

Most of the better quality analog multimeters have a small mirror in the shape of an arc between two of the scales. The purpose of this mirror is to produce a reflection of the needle. When taking a reading position oneself in such a way that the needle is covering it's own reflection. This is then the correct reading. They can be (but not necessarily) less accurate than their digital counterparts. Analog are by far the better meter if there is a fault that results in a slow changing of values within the readings being taken, as the needle will respond to this, whereas the digital meter will just display a jumble of numbers. A down-side to this type of meter is they are less forgiving over polarity. If the leads are clipped on to a power supply the wrong way around, the needle rams itself against the stop. With a digital meter it will just show a negative sign at the side of the voltage reading.

Analog meters are less robust than digital meters. Handle them like a fine clock.

The needle also needs zeroing occasionally, especially after a knock. This is done by using a small screwdriver to turn an adjustment screw that is found somewhere on the body of the multimeter usually near the point where the needle comes through the body of the meter itself. By making very small adjustments to this screw the needle can be returned to its original position to read zero again.

Digital Multimeters

Digital meters can be more expensive to buy, but they are very easy to read. Some will even display a symbol for the test that is being carried out, such as A for amps when testing current, V for volts when testing voltage and Ω for Ohms when testing resistance. These meters are more robust that their analog counterparts (but still treat them with respect).

They give highly accurate readings, good for when a precise reading is needed. They can be a bit of a nuisance on certain types of faults, as mentioned above, as they will just show a jumble of numbers on the display.

They are also very forgiving, for example if a mistake in polarity is made.

Summing Up Which Is the Best?

A good multimeter can last a lifetime if properly looked after. They can save lots of time and money around the home and car. A multimeter can be used for lots of things, from building these projects to sorting out the lights on a Christmas tree. Even if someone has just a passing interest in electronics a multimeter is a must have. As for the best both have their merits, but just for the sheer ease of use the answer must be a digital multimeter (Fig. 4.4).

Another important point to note is that if an unknown value is being measured, for example a voltage, start with the multimeter on its highest range and come down until a reading is found. But always read any instructions given with the multimeter before use. Another good purchase to consider is a small project book explaining how to use a multimeter. These books can be purchased from most electrical suppliers and cost very little. They give far more practical examples of how to connect a multimeter to an electrical circuit then most instructions given with a multimeter. There are several on the market, an example of one is "How to get the most from your Multimeter" by R.A Penfold.

Fig. 4.4 Multimeters digital (*left*) and analog (*right*)

4.4 Electrical Component Identification

When speaking with others about radio astronomy in the past and discussing how it's not all mathematics and big radio dishes, many show real interest, and even ask questions where to obtain the equipment, but as soon as it is mentioned that in some cases one needs to build the receivers and that electrical components are involved, eyes widen and the color drains from faces, as those previously interested take on the look of a rabbit caught in the headlights, and at this point interest quickly evaporates (Fig. 4.5).

Two of the projects covered in this book will require some self-building, namely the INSPIRE receiver and the Radio Jove receiver. If the art of soldering can be mastered and component identification, then building these two receivers will be no more difficult than building a plastic scale model of a 100 pieces or more. No offence to "Airfix" and other providers of scale models, but the construction manuals supplied with the receivers are far better. Plus, once the receivers have been built there is the added bonus of having something that works and can perform a useful task.

It is a nervous feeling after spending several hours building a project when the time comes to switch it on for the first time, but the nerves are quickly forgotten and replaced with feelings of excitement when everything works! The only skill needed is to be able to identify each component, in many cases from a picture or image within the construction manual.

4.4 Electrical Component Identification

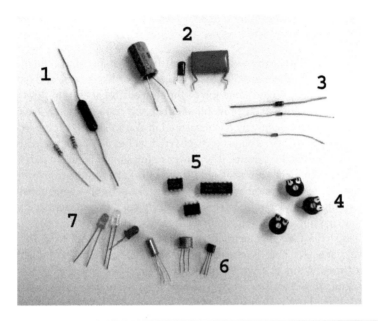

Fig. 4.5 Image showing a range of commonly used electrical components, *1*. resistors, *2*. capacitors, *3*. diodes, *4*. variable resistors and variable capacitors, *5*. integrated circuits (*ICs*), *6*. transistors and *7*. Light emitting diodes (*LEDs*)

The need to know how each component works to build the projects is not necessary, but it can be an advantage to know what job each component does. This can help with the instillation of each part as some can be fitted either way round but others are polarity sensitive and must be fitted in the correct orientation or they will not work. This is especially important in the case of electrolytic capacitors as they can burst open, covering the area in little bits of aluminum foil. Many probably use these electronic components everyday without realizing it. Take for instance the rotary volume control on a car radio, this is nothing more than a variable resistor. Meanwhile, the electronic defribillator paddles which are placed on the chest to deliver the life-saving electric shock to restart the heart is a capacitor that has been charged up and then quickly discharged through the paddles into the body.

The battery charger that is used to charge a car battery, wouldn't work without the help of diodes, that change the alternating current from the power supply into a direct current to charge the battery.

Below is an explanation of the most common components used in electronics and a brief description of the job they do. In order to help anyone wishing to build either of the two receivers the descriptions are accompanied by information on components identification, how to understand the orientation of components and whether the component is polarity sensitive or not.

Resistors

The first component to be discussed is the resistor. Resistors come in many different sizes and shapes, but there are two basic types: one of fixed value and one of variable value. They both do the same job of limiting electrical current flow. Resistors are not polarity sensitive and can be fitted either way round, but they have a power rating associated with them which is measured in Watts. The higher the Wattage the bigger the resistor will be. The projects covered within this book use resistors of a power rating of 0.3 and 0.5 watts. The power rating should be taken into account if choosing to design electronic projects for oneself. For example a favorite trick of the amateur optical astronomer is to use resistors to keep damp moisture from building up on the finder scope. This is done by soldering a number of resistors together and passing a current through them from a battery. As the resistors do their job and resist the flow of electrical current, heat is given off from the resistors themselves, and this heat is enough in some cases to keep the moisture away. All the resistors within the two kits are of the correct power rating.

To illustrate how a resistor works, imagine a garden hose with water coming out of the end. If the hose is then stood on, the water flow will be reduced. This would be equivalent to a resistor in an electrical circuit. How to work out the value of the resistance of a resistor is to use a four colored band code on the resistor itself. The table below outlines how this system works.

Resistor Color Coding

Color	First band	Second band	Multiply	Third band
Black	0	0	×	1
Brown	1	1	×	10
Red	2	2	×	100
Orange	3	3	×	1,000
Yellow	4	4	×	10,000
Green	5	5	×	100,000
Blue	6	6	×	1,000,000
Violet	7	7	×	10,000,000
Gray	8	8	×	100,000,000
White	9	9	×	---------------

The fourth band indicates accuracy to which the resistor is manufactured, to the value shown:

Gold band = within 5 %
Silver band = within 10 %
No forth band within 20 %

For example, if we take a resistor with the four color bands of brown, red, orange and gold. Brown equals the number 1 in the first column, red equals the number 2 in the second column, so we have 12. If we then look at the forth column to get the multiplier we see that it is 1,000. If we then multiply $12 \times 1,000$ we get a value of 12,000 ohms. The fourth color band is the accuracy to which the resistor is manufactured, so a gold band would mean than the resistor is within plus or minus of 5 % of the stated value.

It can take some time to get used to this system, but after building the projects it will become almost second nature. If the colors of a resistor have become faded with age or if there is uncertainty regarding the value, a quick check with a multimeter is all that's needed.

Capacitors

Capacitors come in a wide range of different types, some polarity sensitive and some not, and it is very important not to get this wrong. Capacitors are usually marked in some way if they are polarity sensitive. They can be of fixed value or variable type, like a resistor. Unfortunately, unlike resistors, there is no universal color coding. Much depends on the manufacturer, and while some use a form of color coding other manufacturers stamp the value somewhere on the capacitor itself. The maximum working voltage the capacitor can withstand should also be marked on the capacitor. The value of the fixed capacitor will always be the same even if the voltage that it is used at is lower than the maximum working voltage. Never exceed this maximum working voltage.

The job of the capacitor is to act like a reservoir, but instead of storing water, capacitors store electrons. They can be used in timer circuits in conjunction with a variable resistor, where the variable resistor is used to slow the charging of the capacitor before triggering other actions to take place, such as the intermittent control on a car's windshield wipers. Capacitors can also be used as filters to smooth out spikes within an electrical circuit. Spikes are absorbed within the capacitor itself, in a similar sort of way that the flywheel on a car's engine smoothes out the motion of the pistons to keep the rotation smooth and constant. It is important to remove these spikes, especially with radio receivers, in order not to artificially introduce unwanted noise within the receivers electronics themselves.

Variable capacitors are used in the tuning circuits for radio receivers like the ones covered in this book. They are set by using a small screwdriver, which turns movable plates to alter the size of the storage area. One of these variable capacitors will need to be set after building the Radio Jove receiver.

Capacitors must be respected at all times, as they can still hold an electrical charge for some time after the power has been switched off. The old cathode ray tube televisions are particularly bad for this, as they could hold a charge for days if not longer. Never poke about inside any electrical appliance with anything, least of all a metal screwdriver, as doing so may necessitate the help of a capacitor to restart one's heart. The really dangerous capacitors are the high voltage electrolytic type, having once witnessed a foolish act by an apprentice electrician. He short circuited one of these high voltage electrolytic capacitors to the side of his metal work bench. The capacitor itself must have been the size of a soft drinks can. There was a bright flash and the capacitor's terminals spot welded themselves to the side of his work bench. He had to hacksaw through the capacitor terminals to remove it. The workshop foreman, as can be imagined, wasn't very happy, and neither was the apprentice when he was charged for a replacement capacitor.

The capacitors used in the two projects covered within the book are of low voltage, so no harm should come to those who carefully follow the instructions given within the construction manual provided.

The Diode

Diodes also come in many different types, but in its simplest form a diode's job is to act as a non-return valve in an electrical circuit. Diodes allow electrons to flow one way, and they will block any movement in the opposite direction. It is therefore important that diodes are fitted in the correct orientation. Looking at the casing of a small signal diode, it will be noticed there is a color band at one end. The band denotes the direction of current flow. Remember that the current flows through the diode towards the band, and on observation the correct orientation of the diode will be known.

Diodes come in different power ratings, like resistors, and this power rating must not be exceeded. Looking at a few of the different types of diode that are available will give a better understanding of the different uses of each type. Power rectifying diodes are used to change alternating current into direct current, and they are also used to help control power spikes in high powered switching applications. Depending on the power and voltages that they are being used for, it is not uncommon to find such diodes surrounded with substantial metal heat sinks to dissipate any heat they produce in order to help keep them cool.

Small signal diodes are the sort of diodes that will be used within the projects described in later chapters. They are the little brother of the power rectifying diodes, and while they perform the same sort of job as their big brother, they are much smaller and they can be mounted on a circuit board. Diodes can be used singly or in conjunction with other diodes. If one diode is used to rectify an alternating current into a direct current the diode blocks the reverse current flow, so only half of the alternating current gets through. This is known as half wave rectification. Four diodes can be used in conjunction with each other to form what is known as a "bridge rectifier." This has the advantage of allowing all of the alternating current through, but in such a way that it has now all been changed to direct current. This is known as full wave rectification.

Zener diodes are a special type of diode. Unlike the other diodes discussed above which will block all reverse current flow, zener diodes will, under certain

4.4 Electrical Component Identification

circumstances, allow a reversal of current flow to take place. Different values of voltage can by assigned to the zener diode reversal quality. This means they can be used as a voltage regulator, or a voltage sensitive switch, to protect other components from overloading.

Light emitting diodes or "LEDs" for short, are another special type of diode, but they are made of a different substance to the other diodes described above. This material emits light when an electrical current is passed through it. As with all other diodes they are polarity sensitive and must be fitted in the correct orientation. Looking at an LEDs plastic cap there will be a rim at its base; on closer inspection of this rim it will be noted that there is a flat mark on the rim itself. This denotes the negative connection. LEDs come in a number of different colors and are used in just about every electrical device made, from the standby light on a television set, to the international space station. There are several advantages of LEDs over normal light bulbs. LEDs have a much longer life than a normal light bulb, in some cases 10,000 times as long, and they use far less power than a normal light bulb. This is why lights on a car are slowly being changed to LEDs, and now that the white LED has been developed it can only be imagined what else will be replaced with LEDs. Also an LED is far more robust than a normal light bulb, and unlike a normal light bulb an LED doesn't mind being switched on and off repeatedly. LEDs also operate at a far cooler temperature, so unlike a normal light bulb no burnt fingers when handling them.

Transistors

There are lots of different types of transistors; they come in all different shapes and designs, from small plastic ones right up to the hefty metal-bodied power transistors. In fact, there are whole books listing hundreds, if not thousands, of different types of transistors and listing some two dozen different uses for transistors, such as amplifier, high frequency switching, and power switching, just to name a few. Each transistor is designed to have its own special quality and use. Some are small in size, such as the small signal and switching type used in the construction of the two receivers covered within this book. Others, like power transistors used in high power amplifiers, are quite large and need to be fixed to a heat sink to keep them cool.

Transistors have three wire connections coming from them: a base, collector and emitter. Some transistor designs have only two wire connections, like the high power transistors, and these use the body of the transistor itself as the third connection. The basic principle of operation entails a small current or voltage applied to the base connection, allowing a larger current or voltage to be passed through the collector and emitter connections. Please see image below to see the connections base, collector and emitter shown as B, C and E (other components, such as resistors, have been left out for clarity) (Fig. 4.6).

This means that a transistor can be used as an amplifier. If a small input signal, such as the weak signal picked up from an antenna, is fed into the base of the

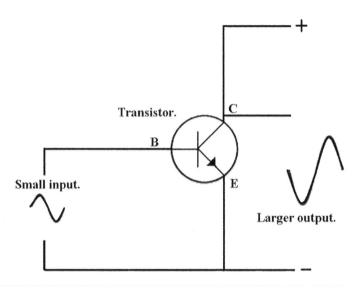

Fig. 4.6 A transistor used as an amplifier

transistor, it will cause a large current to flow through the collector and emitter. This has the effect of amplifying the input signal. If this process is repeated through different stages, the input signal may be amplified up to a level that can eventually be used to drive a speaker. However most of the small signal and switching transistors are quite delicate and can easily be damaged if they are overloaded with too much voltage or current.

It goes without saying that transistors must be fitted in the correct orientation. Some transistors have the connections marked on the body of the transistor itself, but others need data sheets to sort out the connections. The transistors used in the two receivers covered in this book come with very clear instructions and it is unlikely there will be any trouble fitting them in the correct way. Another important point to be taken into consideration is that transistors can be damaged by heat, so it is important when soldering them in place not to leave the soldering iron in contact for more than necessary to solder the connection. It is also a good idea to use a heat shunt to protect the transistor from any unnecessary heating during the soldering process.

Inductors

An inductor, sometimes called a choke, is in its simplest form a coil and can take on a number of different designs, from some that look similar to resistors to wire wound coils. The physical size and shape of an inductor takes into consideration

4.4 Electrical Component Identification 75

two important factors: the power rating and the level of inductance that is required. The inductors used with the two projects covered here have the appearance of a capacitor, but are a dark grey in color, and they are clearly labeled and not polarity sensitive. Without going into too much detail about how inductors work, as an electric current flows through the coil of the inductor a magnetic field is generated. The effect of the magnetic field produced by the current flow is called inductance. Inductors are used in conjunction with capacitors to form tuning circuits, for example the tuning circuit within a radio receiver, or as a filter which will only allow the desired frequencies to enter a circuit.

Integrated Circuits

Integrated circuits, ICs for short, are sometimes called silicon chips or microprocessors. These are the small black rectangular shaped items inside just about every electrical appliance on the planet. Instead of having lots of different components such as transistors, resistors and diodes soldered separately onto a circuit board, an IC has all these components etched on to a single wafer of silicon, hence the name silicon chip. Each wafer of silicon can contain anything from a handful of components to many thousands. The invention of the silicon chip made it possible to make items, such as a computer, from something that filled an entire room to something the size of a briefcase. The ability of having thousands of components on one small wafer of silicon has also increased computing power vastly over the years since its creation.

In the early- to mid-1980s computers such as the Sinclair spectrum ZX 81 were on sale boasting of their 1 kilobyte of memory, but now it is not uncommon for a cell phone to contain an 8 gigabyte memory card that is half the size of a postage stamp. To put this in perspective, when the Apollo 11 crew went to the Moon in July 1969, the computing power they had at their disposal was infinitely less than the computing power of the cell phone we now keep in our pockets. Integrated circuits are supplied in many different types, and they can be used to build timer circuits, amplifiers, and just about anything that can be imagined.

What is needed to build these projects is to know how to orientate an IC to place it on the circuit board the right way up. All ICs, whatever the manufacturer, have the connections numbered and are orientated the same way. Looking at the image below, this shows an eight connection IC known as a "555" that is used, in conjunction with other external parts such as resistors and capacitors, to produce timer circuits. The internal workings of the IC are not important here.

Take note of the small notch at the center. This notch indicates the top of the IC. To the left of this notch a small dot will be seen. This mark can be a white spot or even a small indentation used as a mark. However it is marked, there should be a mark there. This mark shows the position of the number one connection. They are then numbered in the same way as on the image. If the IC has more than eight connections, for example 16, then the number one connection will be at top left then

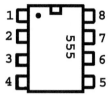

Fig. 4.7 Image of a 555 IC used to show orientation

numbered down to the eighth connection bottom left and then across to bottom right (this will be number 9), back up to the top right to number 16 (Fig. 4.7).

Integrated circuits can be very delicate, and some can even be damaged by the static electricity contained within the human body. An IC of this type should come in a small pack bearing a static electric warning on it. In such circumstances a grounding strap should be worn around the wrist with the other end connected to a suitable earth connection before handling the IC. Thankfully none of these types of ICs will be used here.

ICs can also be very easily damaged by the heat from soldering, even more so than transistors. One way that is used to stop this heat damage is not to solder them. What is used instead is an IC socket, a socket with the right number of connections on it. This socket is then soldered onto the circuit board in place of the IC itself. The IC is then plugged into the socket. This has another advantage that if the IC itself goes wrong for whatever reason, it's just a matter of unplugging it to replace it with a new IC.

There is a very important point to make about plugging in and unplugging an IC from its socket. IC connections are very fragile and can easily be bend or broken, so be very careful when plugging an IC into its socket to make sure that the connections on the IC go into the correct and corresponding connections on the socket. This may seem to be stating the obvious, but the connections on an IC are very close together, and mistakes can easily happen. Sometimes the connections on an IC are spread open too much to fit it in to the socket. If this is the case carefully bend in the connections of the IC so it can be fitted into the socket. A good way to do this is to hold the IC between the finger and the thumb and very carefully push the connections down on top of a table. This will bend all the connections equally, and hopefully keep them all in line. Something that can sometimes turn into a bit of a nightmare is the unplugging of an IC from its socket, especially if it has been in the socket for some time. Luckily this happens only on very rare occasions. It's not a simple matter to pull the IC out, it must be "teased" out of its socket. There are special tools that can be purchased to do this job. Another good way that has been found is to use two very small flat-bladed screwdrivers to fit under each end of the IC, and slowly ease up each end a little at a time in turn in a sort of rocking motion. Don't be too enthusiastic with either screwdriver as the IC connections can bend or even break, or even worse the IC casing could split open.

Treat an IC as if handling a stick of dynamite then nothing should go wrong. In general, to save the risk of accidentally damaging an IC, they should be one of the last components fitted when building up a circuit.

4.5 Headphones

Headphones will be useful for three of the projects covered within this book. Headphones can sometimes be overlooked, but the value of a good pair can't be overstated if only because they can make listening to music a much more pleasurable experience. These need not be expensive. By shopping around it is usually easy to pick up a good pair quite reasonably. Try and avoid the fiddly little ones that come with MP3 players and the like, though some of this type are better than others. Avoid the really cheap headphones as their sound quality is poor and they sound as if they were being used inside a tin can. They can be used with the projects within this book if nothing else is available, but they are really only any good for the use for which they are intended, jogging in the park.

A full size pair of headphones labeled as having a good bass is far better, as they are more comfortable to wear and are not forever falling out. Try to get a pair that have a cushion that fits all the way round the ear, as opposed to those that just sit on the ear. This has two good advantages, it helps keep outside noise out and keeps the noise from the speaker inside. It gives the effect of being in a sound proof room. A good way to test this is to try and listen to some music in a room where the television or radio is switched on using the little ear phones that fit inside the ear. Without changing the volume setting swap to the full size headphones with the cushion that fits around the ear and enter the same room. The sound from the headphones will be clearer and the noise from the television or radio will be less of a distraction.

All the sounds that will be heard from the projects covered in this book will be heard against the constant background hiss of white noise, and the advantage of a good set of headphones can't be stressed enough, as it can make all the difference to hearing a sound or not hearing it. Unlike listening to music where just looking at the CD case will tell what the next track will be, the radio astronomer doesn't have this luxury when sitting and listening for a meteor ping, as they come completely randomly and can happen any time.

What to look for in a good set of headphones is a driver unit, which is basically the size of the speaker in each earpiece, ideally along the lines of 32 millimeters (1.25 inches)–40 millimeters (1.57 inches); a frequency response along the lines of 20 hertz–20 kilohertz as this corresponds to the audio frequency of human hearing; and an inline volume control is handy, but not essential, as it allows the volume to be increased or decreased without touching the receiver. This is good because if the receiver has been calibrated to the computer, as any changes to the receiver volume will mean the calibration procedure will have to be done again.

The length of cable that comes with the headphones may seem unimportant, but it is crucial. One meter (39 inches) is not long enough. The cable will be at full stretch, and the slightest movement could pull the headphones off the head, or worse pull the receiver off the desk and send it crashing to the floor. Three meters (117 inches) is too long, as it would be likely to tie itself in knots and get wrapped around anything and everything. Two meters (78 inches) is just about right, as this length would allow the user to move about without getting the cable wrapped around everything.

Full size headphones usually come with a 3.5 millimeters (0.125 inch) or 6.3 millimeters (0.25 inch) stereo jack plug fitting.

The most common fitting used on MP3 players, cell phones, computers, etc. is the 3.5 millimeters (0.125 inch) jack plug. All the projects covered here require the smaller of the two plugs. Headphones sometimes come with an adaptor that can be used to change one size to the other. The adaptor just plugs on the end either to make the connection to the item larger or smaller. If an adaptor is not included with the headphones they are easily obtainable from most electronics shops for a nominal outlay.

The final, and probably the most important thing, is to remember to make sure they are comfortable. Before purchasing a pair of headphones, ask to try them on. Shoes wouldn't be purchased without trying them on first. If they are too heavy, look around for a lighter weight pair. This is important as they may be worn for many hours at a time. The weight is sometimes stated on the box.

There are also new wireless headphones that are on the market nowadays. These work by having a small transmitter which plugs into the headphone socket of a radio, CD player, etc. This transmitter turns the audio signal into a radio frequency, and then a receiver built into the headphones, turns the radio frequency back into an audio signal which is then played through the earpieces of the headphones, thus eliminating the need for a cable. This may seem ideal, but the thought of having a radio transmitter beaming a radio frequency signal to the headphones in close proximity to a sensitive radio receiver doesn't seem to be a great idea. As interference is the greatest enemy of radio astronomy, it seems sensible not to introduce any unnecessary interference either from the signal, from the transmitting unit, or from the electronic circuitry contained within the transmitter unit itself. Another disadvantage of this type of wireless headphone is that they need batteries to power them, and this will add extra weight to them. Also batteries have a habit of running out at the most inconvenient times. A well chosen pair of headphones need not be expensive and can be used for anything, not just radio astronomy, and if looked after properly can last a lifetime (Fig. 4.8).

4.5 Headphones

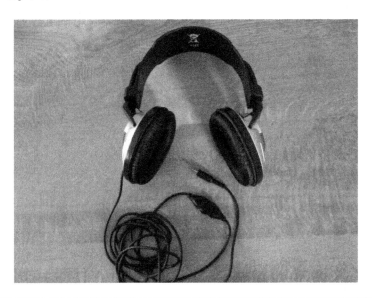

Fig. 4.8 The above image is of a full size pair of light weight headphones with good soundproofing cushions that fits all the way around the ears. There is also an inline volume control. This pair was not expensive to purchase. They are made mostly of plastic, but have soft leather ear cushions and deliver excellent sound quality

Chapter 5

The Stanford Solar Center, SuperSID Monitor

The Stanford Solar Center's SuperSID monitor provides an excellent introduction to the fascinating hobby of radio astronomy, especially for those who do not have much space, as the antenna needed for the SuperSID monitor doesn't need to be placed high up. No antenna mast is needed, and the antenna will work perfectly well at ground level, either indoors or outdoors. In fact an indoor antenna can work very well.

The SuperSID monitor is self-calibrating and specialist knowledge is not required to operate the unit.

The SuperSID monitor can be used to produce real science data or just as an accompaniment to visual astronomy of the Sun. It can be used in conjunction with the Radio Jove receiver, and by looking at the data it gives a good indication that a possible aurora is on the way. The SuperSID monitor is used to monitor the ionosphere for Sudden Ionospheric Disturbances (SID's) and Sudden Enhancement of Signals (SES's). This is done by using the signal from transmitters situated around the world, transmitters that nations use to communicate with their submarines. By monitoring the ionosphere in this way the Sun's activity can be observed in relation to X-ray flares, coronal mass ejections and other phenomena referred to as "space weather."

It is inexpensive to buy and needs minimal DIY skills to build an antenna. The size of the SuperSID monitor is small, no larger than a shoe polish tin, measuring 80 millimeters (3.25 inches) in diameter. It requires no soldering or electronic work at all, except for the fitting of a screw on Bayonet Neill-Concelman or BNC connector for short (instructions are supplied for this) and the coupling of wires into a jointing box using a screwdriver. Please see image (Fig. 5.1).

Fig. 5.1 The SuperSID monitor

The SuperSID monitor comes with its own power supply, in the form of a transformer. If the monitor is being used in the United Kingdom an adaptor will be needed; a two pin electric shaver plug is just fine for this purpose. It also comes with the required fittings, although it will require a short length of RG 58 50 ohm antenna cable. This is available from any electronic shop. Wire will be needed to make the antenna itself, a minimum of 120 meters (393 feet) of 22 AWG (American Wire Gauge) or 23 SWG (Standard Wire Gauge) will be fine, this too can be got from an electronic shop. This wire will then be wrapped around the antenna frame after its construction, to form a working antenna for the SuperSID monitor. The construction of a suitable antenna will be covered later in the chapter.

Ordering the kit over the internet is easy, from either the Stanford Solar Center or from the Society of Amateur Radio Astronomers (SARA) website. It arrived promptly within a couple of weeks. If ordering for delivery into the United Kingdom or elsewhere there is the usual import duty to pay, and extra postal charges will apply. Included with the SuperSID monitor will be a CD-ROM with the operating program and other information such as a printable manual. It is handy to have a hard copy of the manual to browse through. There is also a printed sheet with one's own personal monitor ID number, plus the site name that was chosen before ordering. Keep this information safe as it will be needed later. This information is used if wishing to take part in research, as the program can be set to automatically upload the SuperSID data back to the Stanford Solar Center, where it is collected and processed from observers around the world. This is optional, and everything will work just as well if choosing to be a stand-alone observer.

Fig. 5.2 SuperSID parcel unpacked with other space goodies

There is also an email address and in case any problems arise, a member from the Stanford Solar Center will be able to answer any questions or offer advice. They are very quick when it comes to replying and in most cases an answer is received the following day.

There are a number of other space "goodies" included with the SuperSID monitor these are a NASA Sun fridge magnet, a 3D post card of the Sun, a 3D NASA book mark, a large 3D poster of the Sun, and a DVD, entitled "Cosmic Collisions", along with leaflets and a sticker for SARA (Society of Amateur Radio Astronomers). Please see image (Fig. 5.2).

Once the SuperSID monitor is up and running no further input is needed, apart from checking the condition of the antenna every now and then, if it is kept outside.

The monitor is quite happy just doing its own thing recording data. All the user need do is check the data for evidence of solar activity. Using the SuperSID monitor to observe the Sun every day will also lead to changes being noticed in the ionosphere directly overhead. These changes can be seen from month to month and season to season. Once familiar with understanding the data, this project is very easy and very rewarding to use. 2013–2018 would be an excellent time to purchase a SuperSID monitor as the Sun is building up to the solar maximum again, and between these years the Sun should be producing some useful data. Although it is not unheard of for the Sun to throw out the odd surprise, like a massive X-ray flare during solar minimum without any warning whatsoever.

5.1 Space Weather and Its Dangers to Earth

The Sun has been throwing out a constant stream of charged particles, known as the solar wind, ever since the first day that nuclear fusion started within its core millions of years ago. The intensity of the solar wind has a lot to do with the 11-year solar cycle, where the polarity of the Sun's magnetic field flips. At solar minimum the magnetic field lines enter the magnetic poles of the Sun, these magnetic poles are close to the axis of rotation of the Sun. Midway between the flip is what astronomers call the solar maximum. The magnetic field lines are now at right angles to the Sun's axis of rotation. With the Sun being a gaseous body and having no solid surface, it rotates at different speeds at the equator and the poles. This has the effect of twisting and distorting the magnetic field lines. For example imagine a garden hose that instead of being neatly coiled on a reel is just thrown in a heap; the hose will soon become kinked and knotted. The tension builds up in the field lines like giant elastic bands and at some stage something has to give. The magnetic field lines snap and can break through the surface and produce sunspots and other solar outbursts. Sunspots sometimes come in pairs, with one of a north polarity and the other of a south polarity, just like a magnet.

When magnetic field lines break, a large amount of energy is released, and this is known as a coronal mass ejection, sometimes abbreviated to CME. This energy travels through space in two particular forms, one as a high energy electromagnetic wave travelling at the speed of light that reaches the Earth in a little over 8 minutes, and the other as high energy charged particles within the solar wind that can travel at speeds of several thousand miles per hour. Depending on their velocity, these can take somewhere in the region of 2 or 3 days to reach the Earth.

This type of solar activity has been happening for millions of years, so why bother about it now? The reason is that any manned space flights carrying astronauts in space, for example returning to the Moon, will be in trouble if one of these huge CME's explodes on the Sun, as all they will have to protect them is the spacecraft itself. In a little over 8 minutes they will be exposed to huge amounts of lethal doses of extreme ultraviolet radiation and X-rays equivalent to thousands of medical X-rays. If they somehow managed to survive this, in a day or so they would be hit by the high energy partials within the solar wind that would have so much energy that they could pass straight through the spacecraft including the astronauts inside. The risks would be even greater with a manned mission to Mars, as the round trip would be in the region of 3 years.

It isn't practical to carry the lead screening needed to protect the astronauts against these types of radiation, and the best idea that scientists have come up with at the moment is to use the onboard water as a protective screen. This would be achieved by having a room in the center of the spacecraft with the water surrounding it. The crew would have to stay within the confines of this one room until the threat overtakes the spacecraft. Depending on the distance at which the crew were from the Earth, the type of threat involved and the speed at which it was travelling, at best we could give the crew maybe a couple of days warning in which to prepare and take the necessary precautions open to them.

Here on Earth we can also be at the mercy of high energy radiation through our reliance on satellites. Satellites, in orbit above the protective atmosphere, can be subjected to high energy radiation from the Sun. These include the global positioning satellites and the cell phone network of satellites. The high energy radiation can overload their delicate electronics and literally "fry" them just like placing a cell phone in a microwave oven. This could lead to all manner of problems with navigation and communication, and could prove very expensive for businesses as international banking, for example, uses satellites.

There is another problem that can affect thousands of people and leave them without the ability to heat their homes or cook a meal. All over the world we have high voltage power lines that carry electricity from the power stations to our homes; they can stretch for hundreds of miles. These power lines can act like a huge antenna and collect the high energy radiation, and this can induce an electrical current to flow within the wires of the power lines causing them to overheat, or even worse to overload. This can be a real problem for the transformers that step the high voltage down again for use in our homes, as these can overload and burn out. The only option that is open to the power company is to shut down the area that is likely to be affected, but this has its own problems. The process is not as simple as just turning off a light switch. This is not too much of a problem for hydro-electric generation, where valves can be shut to cut off the flow of water, but for turbines using steam to generate the electricity these turbines can take days to slow down before they can be stopped, so the power needs to be diverted away from the affected area. There is the added problem of vital services like hospitals and the other emergency services, not to mention the domestic consumers who would not be too happy being without electricity.

All these services rely on accurate space weather forecasting. A warning is normally issued when X-ray solar activity with a magnitude of "M 5" is detected. (X-ray classification will be covered later in the chapter). So it can be seen that there is a genuine threat from the high energy radiation from the Sun. SOHO (Solar and Heliospheric Observatory) can give us some warning that something is on the way, and to try and take the best course of action with the information available. But SOHO can only see the side of the Sun that faces the Earth. So if there were a large amount of activity round on the side of the Sun which SOHO couldn't see, it would be unknown to us until the Sun's rotation brought it into view. The STEREO mission (Solar TErrestrial RElations Observatory) consisting of twin spacecrafts positioned on each side of the Sun in such a way that between them they can see all of the Sun compensates for this blindspot. Due to the position of the spacecrafts, if solar activity were building up on the side facing away from the Earth we have more of a warning that something may be about to happen.

5.2 A Basic Description of How the SuperSID Monitor Works

Situated around the globe of the Earth are powerful transmitters that different countries use to communicate with their submarines. The antennas for these transmitters are huge, measured in kilometers, due to the wavelength at which they operate.

As we have already discussed, within the properties of longer wavelengths is its ability to penetrate objects such as the deepest oceans of the world. These transmitters are so powerful that the signals from them can travel around the Earth, bouncing off the ionosphere. These frequencies have the advantage of being very stable. The SuperSID monitor doesn't receive the communications like a radio receives music, no communication between the submarines and the bases on land, can be heard. Any communication would be in code anyway for obvious reasons of security. We are only interested in the carrier wave frequency from these transmitters.

A carrier wave in its simplest form is a signal frequency wave that is used to carry information, and can be modulated by either amplitude or frequency, and now digitally coded.

One way to describe this is to imagine a radio in a car tuned to a station; the music is the modulation or change within the signal, but the frequency at which the radio is set is still the same. The SuperSID monitor is only interested in this carrier wave.

The signal from these powerful transmitters bounces between the Earth and the "D" layer of the ionosphere, and by doing this it can travel around the world. The SuperSID monitor picks up these signals through its antenna and amplifies them to give a useable signal for recording and processing on a computer. If there is any change in the ionosphere, for instance as the result of a solar X-ray flare from the Sun, this can cause a sudden drop in the signal strength. These are called SID's as they arise from Sudden Ionospheric Disturbances. Solar flare activity can also cause a sudden increase of signal strength, and these are referred to as SES's for Sudden Enhancement of Signals. The SuperSID monitor will pick up this change in signal strength. This is only possible between the observer's local sunrise and sunset times, as the ionization in the "D" layer is lost at night and the monitor, can no longer function without the ionized "D" layer. This means only the data from local sunrise to sunset needs to be checked.

If there is a large solar flare, the monitor can sometimes be overwhelmed and the signal maybe lost completely. This doesn't harm the monitor, but it will show up on the graph when processed later. Sometimes a transmitter signal will be lost for a short time, and the graph will just show a flat line. If there has been no solar activity to overwhelm the monitor this could indicate that the transmitter, has probably been shut down for a short time, in order for maintenance work to be carried out. It has been suggested that by taking note of the time and date of the transmitter shut down, a pattern of maintenance of the transmitter can sometimes be worked out. A list of suitable transmitters can be found within the CD-ROM supplied with the SuperSID monitor, or from the Stanford Solar Center website. Each observer will choose their own transmitter(s) on the basis of signal strength, and this process will be covered in a later chapter.

The SuperSID monitor is designed to monitor six different transmitters simultaneously, and it takes a reading every 5 seconds throughout the day. At the end of a day, using the software provided, a graph can be produced, and this graph can then be examined for traces of solar activity. Looking at the image Fig. 5.3, this is the sort of thing to watch out for. This is a portion of a graph starting at 8 am and ending

5.2 A Basic Description of How the SuperSID Monitor Works

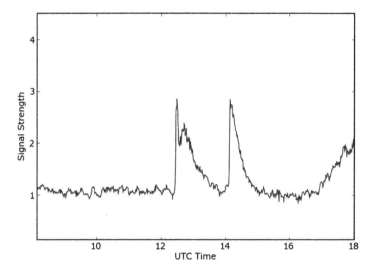

Fig. 5.3 SuperSID data showing two flares

at 6 pm showing two changes within the signal that the SuperSID monitor has recorded (Fig. 5.3).

These are X-ray flares and will be covered more in-depth later in the chapter, but for now look at the peaks the line on the left of each peak is vertical. This shows that the ionization of the ionosphere happened almost instantly, while the right-hand side shows that the ionization decayed to background levels over almost 2 hours for the first peak and nearly an hour for the second. This image is the classic shape to watch out for, although the peaks could go down as well as up depending on whether it is SID or SES change in the signal. The height of the peak can change depending on the intensity of the flare, and some flares can last an hour or so, like the ones above, while others can last all day.

For those who are interested in using the SuperSID monitor with other forms of solar activity, it a good idea to sign-up to an aurora warning service. There are plenty available and a quick search online will show a number of them. It is important to sign-up to a local one or at least one within the local time zone, as this makes information that is sent more relevant. These aurora warning services automatically send out emails to let recipients know that there is a high probability that an aurora may be observed over the next couple of nights.

If a large peak is picked up within the SuperSID data the recipient should keep an eye on their emails because it is likely to trigger an aurora warning to be issued. This is due to the fact that at the time an X-ray flare is released from the Sun, a vast increase in the high energy particles which cause the Earth's auroral display will also be released. As X-ray flares burst from the Sun with tremendous power the electromagnetic energy from these flares are travelling at the speed of light and reach the Earth and our SuperSID monitors in a little over 8 minutes.

Although the high energy particles which are released at the same time as the flare travels at a slower rate of several hundred miles per hour, they can take up to 2 or 3 days to reach the Earth before putting on a display of aurora. It is good if the X-ray flare has been captured on the SuperSID equipment and especially good if an aurora is observed a day or so later, as this completes the picture. As this shows that there really is a link between X-ray flares and the interaction of the high energy particles and the Earth's magnetic field, and that the SuperSID equipment has detected this.

If a suspected flare has been captured, this can be checked at the NASA 3-day X-ray web page. This is updated every few minutes and it can be easily checked for X-ray solar activity. Simply check the time of the suspected flare against the time on the NASA web page to confirm it. Before too long these flares become easy to recognize within the SuperSID data. It is useful checking for the smallest flare that can be picked up, this will show how sensitive one's equipment is, and how well it performs.

Lightning can sometimes cause spikes on the data, but these can be spotted quite easily with practice, especially if the weather forecast has predicted thunder storms, although thunder storms don't have to be local events for the SuperSID monitor to pick them up.

As mentioned earlier, over the space of a year changes will be noticed within the graphs depending on whether or not there is solar activity. A dramatic change can be seen from midsummer to midwinter, caused by the fact that the Sun changes it's altitude in the sky. These changes will be unique to each observing site, as any change in latitude of the Sun will make all the difference to the equipment and how the data is recorded.

5.3 How to Make an Antenna

The antenna for a SuperSID monitor is a simple wire wound antenna, and can be made from a variety of materials such as wood, plastic or a mixture of the two. The shape of the antenna isn't of too much importance either, the simplest design is a wooden cross. Please see image (Fig. 5.4).

The image shown has an antenna that has four spokes, and each spoke is 705 millimeters (27.8 inches) in length when measured from the center. This gives a total area of 1 meter squared. The 120 meters (393.7 feet) of 22 AWG 23 SWG wire is then wrapped around the edges to form a square measuring 1 meter (39.37 inches) on each side. Solid core wire is preferred to stranded core wire, as the solid core wire holds its shape better because it is less flexible than stranded wire. Stranded wire can be used, but it would need more supporting. This is the smallest recommended antenna size and possibly the easiest to make. It doesn't have to be square, and it can have six, seven and even eight sides, or it could be circular. It all boils down to the room available to place the antenna and practical DIY skills.

The size of the antenna has a lot to do with location. For those living in a location with a high strength signal this small antenna would work fine. But on the other hand a larger antenna would give more gain and would possibly provide more

5.3 How to Make an Antenna

Fig. 5.4 A simple SuperSID antenna

stations to monitor. The pictured antenna was the first antenna built for the SuperSID project; it had 200 meters (656 feet) of wire wrapped in 50 turns around the frame, and this gave a total area of 50 square meters. It is better to have 25 turns on a 2 meter (72 inch) frame than 50 turns on a 1 meter (39.37 inches) frame. A 2 meter (39. 37 inches) frame with 25 turns would give an area of 100 square meters, four times the size of a 1 meter antenna.

This first antenna produced a signal, but not one of a suitable strength to use. Within the SuperSID manual it is recommended that adding more turns of wire will improve the sensitivity, so another 100 meters (328 feet) was added. This was an extra 25 turns to the frame which gave a total of 75 square meters, and although this helped, the signal strength still wasn't high enough. So an additional 100 meters (328 feet) were added, making 400 meters in total, with 100 turns of the antenna and 100 square meters in area. This created a problem as the sensitivity had increased but the resistance within the wire had also increased, so any benefit from the added wire was cancelled out by the resistance within the wire itself. The only option was to build a larger antenna frame. Please see image (Fig. 5.5).

This particular shape was used because it gave more area than the smaller square antenna and also followed the shape of the inside of the roof. Each spoke is 750 millimeters (29.5 inches) when measured from the center, and this gave a total area of 75 square meters using only 200 meters (656 feet) of wire. This showed a marked improvement to the signal being received, and was enough for the SuperSID monitor to function correctly. As can be seen from the image, the antenna was the largest that could be accommodated in the location where it was to be kept. But larger antennas could be used if there is room.

Fig. 5.5 Improved SuperSID antenna. Situated inside attic space

To make a simple 2 meter square antenna the following will be needed: two pieces of wood 50 millimeters (2 inches) square section, 2.8 meters (111 inches) in length, an 8 millimeters (0.31 inch) bolt, some 125 millimeters (5 inch) in length, assorted screws, electrical tape, wood glue, a small amount of 6 millimeters (0.25 inch) plywood to support the center, and enough wood to make a simple stand to support the finished antenna, plus wood preservative and or paint if the antenna is to be kept outside. Start by cutting a joint in the center of the two lengths that will make the frame of the antenna. Please see image (Fig. 5.6).

This will allow both pieces to fit at 90 degrees to each other. Try and keep this joint as tight a fit as possible because this is going to take the weight when winding the antenna. Secondly, cut a slot in the end of each of the four ends to accept the wire. Please see image (Fig. 5.7).

When these have been cut, a good tip to save problems later is to round off the inside of the slots so that the wire fits round a curve rather than a sharp edge. This will save the wire from suffering any undue stress, especially if the antenna is to be kept outdoors where movement by the wind could cause the wire to break. Please see image (Fig. 5.8).

5.3 How to Make an Antenna

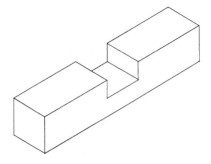

Fig. 5.6 Central joint of frame (not to scale)

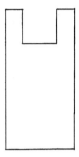

Fig. 5.7 Slot cut ready for wire (not to scale)

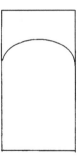

Fig. 5.8 Inside of slot rounded off to protect wire from sharp corners. (Not to scale)

Now it is time to assemble the antenna. First glue and screw the two lengths together at the central joint, the screw is only there to hold the joint together while the glue dries, and it is best to leave this overnight to dry. Then cut two pieces of 6 millimeters (0.25 inch) plywood into pieces 300 millimeters (12 inch) square, to be used to support the center of the frame. Remove the center screw when the glue

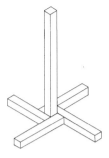

Fig. 5.9 A simple design for an antenna stand (not to scale)

has dried and place a square of ply each side of the frame over the center for support (See the image of the antenna in the attic), glue and screw these on.

Cut a small piece of 6 millimeters (0.25 inch) plywood approximately 50 millimeters×75 millimeters (2 inch×3 inch), and screw this to one of the spokes about 75 millimeters (3 inch) from the end. This is for the jointing box for the antenna cable. Build a stand to support the antenna. This can be any design see image Fig. 5.9. This is a simple design, and no joints need to be cut. This is the type used in the images above and can be easily made using the same size material used for the antenna frame. The height of the stand need only be just over half the size of the finished antenna, and the 4 feet would be quite adequate to hold the weight of the antenna. If leaving the antenna outside permanently a more solid support will be needed, or spikes could be driven into the ground through the 4 feet, after first checking for power and telephone cables, gas or water pipes (Fig. 5.9).

When everything is dry, clean off any rough edges with sandpaper, and apply the paint, varnish or wood protection of choice. It is now time to wind the antenna. Start by drilling through the center of the antenna frame to accept the 8 millimeters (0.31 inch) bolt. Lift the antenna frame against the stand and leave enough clearance so that the antenna can be turned without catching the bottom of the stand. Drill a corresponding hole in the frame support. Now, slide the bolt through the frame, then slide 2 washers on to the bolt and lift the antenna frame into place. Slide another washer on the bolt and loosely tighten the nut. It should now look like a rather crude windmill with its blades. Fix the jointing box to the frame ready for fixing the wire and antenna cable. The wire in the United Kingdom comes on 100 meter (109 yard) reels, and two reels are needed. Join one end of the wire from one reel to the other, preferably by soldering and then covering the joint with electrical tape. Now wind both reels on to one. There should be room on one reel for all the wire. Cut several lengths of electrical tape into 50 millimeters (2 inch) lengths and stick these somewhere handy, like the edge of a table, as this will save struggling later. Eyelets for the jointing box are supplied. Fix one to the end of the wire either by crimping, or better still by soldering.

We are now ready to wind the antenna. First couple the eyelet to the jointing box, leaving a little slack wire, and apply a little electrical tape to the antenna frame

5.3 How to Make an Antenna

to hold it in place. Ask an assistant to help with the winding by holding the reel of wire, and applying a light pressure to act as a brake to stop the wire from becoming knotted or tangled, avoid kinking the wire, as this will weaken it.

Start to turn the antenna and lay the wire in the slots at the end of the spokes. Keep a steady tension on the wire, as this will help it support itself. Apply a little electrical tape every now and then to help keep the wire in place, and carry on until the end of the reel is reached. Take it to the nearest full turn back to the spoke with the jointing box. Leaving enough wire to make the connection, wrap electrical tape around the spoke to hold the wire from coming undone, and this concludes the antenna winding.

Now fix the other eyelet to the end of the wire and fix to the other connection on the jointing box. If a multimeter is available, set it for resistance and touch the two ends of the wire. If the reading is about 10 ohm or less then the antenna is fine, as this is just the resistance within the wire. If there is no reading then this means there is a fault somewhere, like a poor connection or broken wire. Assuming that everything is fine, put the spoke with the jointing box at the bottom and tighten the center bolt. It is a good idea to drill through the antenna support into one of the spokes and fit a screw to hold the antenna solid and stop it rotating and damaging the cable. A spacer will be needed to take up the space of the plywood supporting the center of the frame, and this can be done with washers or an off cut of the same plywood used earlier.

Next, fix the coaxial antenna cable to the jointing box, before stripping the insulation off the antenna cable it is an idea to fit two or three toroid collars to the end of the cable that is coupled to the antenna itself. These just slide on the cable and after the connection is made they are slid back up as close to the connection as possible and held in place with electrical tape or a cable tie. The idea behind these collars is that they help stop interference involving radio frequencies entering through the open end of the antenna cable. This innovation improves the antennas performance, and they can be used on almost any antenna. They are easily purchased from electronic suppliers. They come in a range of sizes, so try to get the ones that are the closest match to the diameter of antenna cable being used.

Start by stripping off the outer insulation of the cable, being careful not to damage the braiding below. One or two broken strands of braiding are inevitable, as long as it doesn't run to half a dozen or more. Carefully, using a small screw driver or something similar, untangle the braiding, it's a good idea to use the tip of a ball point pen for this, as there are no sharp edges. When this is done, twist it together to form a single wire. Next, strip the insulation off the center conductor, but no more than needed to make a connection. As before be careful not to damage or score the central wires. Crimp or solder on the other two eyelets and couple on to the jointing box, it doesn't matter which way round they are. Leaving a little slack cable, apply a plastic cable tie or electrical tape to the cable to take the weight off the connections. If the antenna is going to be used outside be sure to apply enough electrical tape or other waterproofing to protect the connections, waterproofing of antenna cable is covered in the Radio Jove chapter.

Next, fit the BNC connector. To do this, mark off 20 millimeters (0.75 inch) of the outer insulation, being careful as before not to damage the braiding below.

Fold the braiding back over itself. Then strip about 6 millimeters (0.25 inch) off the central insulator and make sure the central conductor has no stray wires. Twist any stray together.

Now, turning the BNC connector clockwise, twist it over the end of the antenna cable. It should only need hand pressure to fit it, so don't be tempted to use pliers to force it on. After the BNC connector is fitted it's a good idea to wrap one or two turns of electrical tape around the spot where the BNC connector meets the antenna cables outer insulation, as an added precaution to keep out damp and dirt.

Try and keep the cable from the antenna to the SuperSID monitor as short as possible, as this makes a difference to the signal. The recommended length is 10 meters, but if 8 meters will do then use that. The antenna is now ready for use. For more ideas about different antenna designs visit the Stanford Solar Center website to explore all manner of innovative and clever ideas for antenna designs.

5.4 Placing the SuperSID Antenna

Now that the antenna has been made it must be located in a place where it can perform its duties. This doesn't have to be high up, as the antenna will work at ground level. First the obligatory safety warnings; if a few common sense precautions are followed then no harm should arise. The antenna is likely to be bulky and heavy, depending on the design, and assistance may be required to maneuver it. Don't mount the antenna anywhere it is likely to cause an obstruction. Check that the antenna cable cannot be tripped over. If mounting the antenna indoors, make sure to take account of the locations of any power cables or water and gas pipes before drilling into a wall. If working at height, for instance off a ladder, be sure to ask for assistance from a reliable person. If the antenna is to be mounted outside make sure any fittings are secure and there is no danger of the antenna falling, especially if it is windy. Don't fix the antenna higher than surrounding buildings or trees because of the risk of a lightning strike. In fact, in the event of a local thunder storm unplug the antenna from the SuperSID monitor until the storm passes.

The antenna must be placed as far away from electrical interference as possible. This can be easier said than done as electrical interference is likely to be everywhere. It is more of a case of finding a location with a tolerable level rather than finding an interference-free location. Some questions to address, if the antenna is being used indoors include the following: Is the building a wooden or brick structure? Does the building contain a large amount of metal or a metal frame structure? If using the antenna inside a wooden or brick structure it should work fine, but if inside a metal building or metal framed building there is likely to be a problem of interference from the metal structure.

Some people have reported problems with interference from outside lighting, in particular from street lighting. This hasn't been a problem here in the United Kingdom as the low pressure sodium lights are being slowly changed to the more efficient high pressure sodium lights. These also cast the light downward and not skyward (not before time). It could be that interference from other sources have overshadowed or masked this problem. The old type computer monitors can be a problem, but the newer TFT monitors on laptops, etc. seem to pose less of an interference problem. It is advised within the SuperSID manual to switch off the older type computer monitors whilst collecting data with the SuperSID monitor, if the computer monitor is a separate unit.

The old cathode ray tube type of television can also be a problem, so try and get as far away from these as possible. Florescent lights are another source sometimes reported as a problem, especially the older ones, but more modern fittings have reduced this. Touch sensitive lights, the type that come on at different levels of brightness and normally used for bedside lights, also can cause a certain level of interference. It is much better to have the simple on/off switch type. Microwave ovens are a real problem, so be sure to keep the antenna as far away from them as possible. One further item that has proved to be a real problem is the computer printer. When printing it drives the SuperSID monitor's display wild. This is easily sorted out by only using the printer after sunset. If there's an interference problem and its origin is not known, switch off every electrical appliance in the home, and then switch on one at a time and take note of the display. This may sound a little over the top and time consuming but it works. By using this method it was found that the television set was such a problem.

All that said, it is still possible to run a successful setup indoors as long as potential problems are kept in mind. The first plan for the SuperSID monitor was to keep the antenna in the spare bedroom with the laptop, but this idea was dashed thanks to the downstairs television set and the microwave oven in the kitchen (which is directly below the spare bedroom). By simply transferring the antenna to the attic and drilling a small hole through the ceiling for the antenna cable, this put more distance between the offending television set and microwave oven, which allowed the SuperSID monitor to work correctly. This had the added advantage that by being in the attic the antenna is less likely to be knocked or moved. The antenna was also briefly tried outside before taking up residence in the attic, but it just picked up interference from the surrounding houses. Everyone's situations and locations are different, but by persevering, a point where the antenna will deliver a working signal can usually be found.

5.5 Computer Sound Cards and Software

The SuperSID monitor comes with its own operating program on a CD-ROM. The specification of computer needed to operate the program are Windows 2000 or more recent and the processor speed must be 1 gigahertz or faster with 128 megabyte Ram or more and depending where in the world one lives a HD sound card with a sample rate of 96 kilohertz.

Sound Cards

The antenna detects the very small signals and then these are amplified by the SuperSID monitor to produce a signal of a suitable strength to work with. These are then fed into the sound card on the computer. The job of the sound card is to detect the information within the inputted signal, and process this into useful information that the operating program can use and store digitally, for inspection and processing later. Any change in signal strength will be detected and registered as a possible SID or SES event. The type of signals used come from the communication transmitters that different nations use to communicate with their submarines. These are very powerful transmitters with many megawatts of transmitting power, but depending where in the world the SuperSID monitor is used some stations will be of more use than others. The sound card allows a greater number of stations and frequencies to be monitored simultaneously, meaning there is a better chance of picking up changes within the ionosphere through any changes detected in the signals. Those living in Europe or Asia can use a sound card rated at 48 kilohertz, but those living in America will need a sound card rated at 96 kilohertz in order to pick up the stations.

To find out the details of a sound card, click on "my computer" tab, then click the "open control panel" tab, look for "audio manager" and click this tab and the details of the sound card should be displayed. For desktop computers with a tower unit it is possible to change the sound card for one with a higher sampling rate. When ordering the SuperSID monitor it is also possible to purchase a HD sound card at the same time from the SuperSID monitor supplier. Very clear instructions are supplied within the printable instruction manual, and these cover fitting the sound card into the tower unit. This however is becoming less of a problem now because as computer technology is increasing over the years and processor speeds get faster the quality of the sound cards have increased along with everything else. The laptop being used to write this was manufactured around 2007 and it contains an HD sound card of 96 kilohertz sampling rate, so it would be a fair assumption to assume that anything younger than this will probably already have a 96 kilohertz HD sound card already installed, but it is worth just checking first.

Software

There is one point that needs to be mentioned before the discussion of the software. There have been reports that Windows Vista has been a problem. It must be stressed that this has nothing what to do with the supplied software, but with the computers themselves. The program runs perfectly well once the program is loaded, the problem is getting the computer to recognize the CD-ROM. This has happened several times over the years with other CD-ROMs, some will load and some won't. Every attempt at trying to load the CD-ROM on to the computer, would cause it to put up a window asking permission to format the blank disc. Should this happen, press

5.5 Computer Sound Cards and Software

"cancel" immediately and remove the disc, or the computer may overwrite (erase) all of the data on the disc. The only way that could be found to get the computer to recognize the CD-ROM was to load it onto the computer by using an old external powered CD-rewriter. The computer was more than happy to load the program from this and once the program was loaded there has been no further problems. This is solely due to the computer and not the fault in anyway of the CD-ROM supplied with the SuperSID monitor. This is the only problem with the SuperSID software of which has been reported, and it seems to be limited to the Windows Vista operating program.

Before loading the program have the paper work that comes with the SuperSID monitor at hand, as this information contains the monitor number and the site name chosen before ordering. Knowing one's latitude and longitude is also required, and this can be worked out from a map or nowadays switching on the GPS device in the car or even looking at a cell phone. The time zone in relation to UT (Universal time) and the sampling rate at which the sound card will be operating at on the chosen computer also need to be known.

Insert the disc with the operating program into the computer and it will probably start automatically. If it does not, then right click on the disc drive short cut and click "open". The software needs the above information before the program is run for the first time. Click on the "configure file" icon. This will open the "parameters" window, and this is where the specific information of the chosen site, etc. must be entered. The window displays in a basic note pad format, and the only changes required are: the site name, longitude and latitude, time zone information, monitor ID number and sound card sampling rate. At this stage leave everything else at their default settings. This may sound difficult, but it is no harder than programming a destination into the satellite navigation device in the car. It's a good idea to print off the pages of the instruction manual that relate to the installation of the operating program first, as this provides a visual guide that is handy to refer to.

At this stage station frequencies will not be known and can be left at default settings for the moment. The automatic upload setting, which sends SuperSID data back to Stanford, is set for "no" by default. If choosing to send off information later on, it's a simple matter of just clicking this box to change it. To start the program it is necessary to access the SuperSID file and click on "superSID.exe", although its a lot easier just to put a short cut to this file on the desktop and just click this. It just saves time. This can also be done with the configure file, so this too can be accessed easier, to change station information and set frequencies when the best station for the location have been found. This will be covered in the next section. If all goes well, and the installation is a success, double click on the short cut icon to start the operating program it should look something like the screen shot below. Please note the shape of the graph, although this is far from an ideal noise-free graph it should give a general idea of what to look for (Fig. 5.10).

Another point worth mentioning about the program is that it is quite happy doing its own thing in the background as it doesn't interfere with the other programs on the computer. Just drop it down on the tool bar and use the computer as normal, except for using a printer.

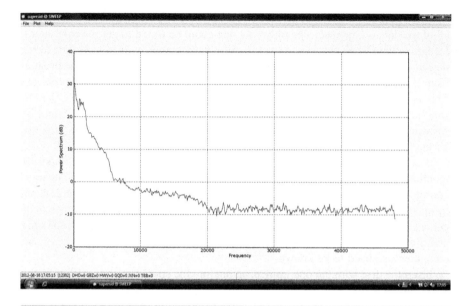

Fig. 5.10 Screen shot of SuperSID's operating program after the software has been installed

5.6 Connecting Everything Together

The time has come to link all the parts together. Connect the antenna via the BNC connector to the SuperSID monitor. This is a simple bayonet fitting, so a push and twist clockwise movement will do the trick. Remembering to keep this cable as short as possible, but not to have it at full stretch, or the cable will suffer from undue stress. It was found that removing a short length of cable, such as 1 meter (39 inches), can make a lot of difference to the signal. Next, plug the SuperSID monitor into a power supply. An adaptor may be needed for this, depending on the design of the domestic electric power sockets. The next part is to plug the "jack" plug into the computer. Don't be tempted to use any type of extension cable such as the sort used to extend the cable of a pair of headphones, as it won't work and the power of the signal will be lost.

Coupling things to a computer can be a little bit of a minefield as each computer will have its own fittings and operating system. So the best way to describe this is to use a familiar computer as a guide. Plug the jack plug into the inline input socket if it has a separate one, but note that this can sometimes switch on the built-in microphone. If this is the case it will be necessary to access the audio properties on the computer and check the box for the microphone input, or the microphone will be live and record all background sound. Some computers share the same input socket for inline and microphone input. When plugging anything into the socket, an on-screen window pops up to ask which input either microphone or inline input is required.

5.6 Connecting Everything Together

The computer may ask what sample rate is required from the sound card's operation.

If there are any problems with the SuperSID's operating program malfunctioning, check that the sound card sample rate matches the sample rate that is set within the "Parameter window", as discussed earlier, this will in most cases probably sort out any problem.

If there is no inline input on the computer, the microphone input can be used, but this is a far more sensitive input and caution will be needed to set the volume input level, as this could cause problems. If the input volume is set too high it will cause distortion of the input, in a similar way to a speaker distorting music played through it if the volume is turned up too high. If the microphone input has to be used experiment with the volume control in order to find a suitable setting. Assuming everything has gone to plan so far, the next thing is to start the program.

Double click on the shortcut and wait a second or two and a screen will appear. In the bottom left hand corner "Waiting for timer…" will be seen, and a second or two later the graph will appear. If lucky some large narrow peaks will be seen. If not, don't worry. These peaks can be one of two things: interference which we don't want, for example from a television set, or the stations that we do want. At this stage it will not be possible to tell which is which. If lucky enough to have a peak, then place the computer curser at the top of the peak and left click the mouse button. In the lower left of the screen three important pieces of information will be shown: frequency, power, and strength. Doing the same thing on a different peak the frequency, power, and strength will have changed. This is how to find out what frequencies give the signals and therefore what stations can be received. These can then be set in the parameter window.

If there are no peaks, or even if there are, it is a good idea to turn the antenna to find the best reception. This is done by turning the antenna about its vertical axis and at the same time looking for the tallest peaks above the background level of noise. This is where the help of an assistant can be useful. Turn the antenna about its vertical axis but by only a few degrees at a time. It is very important to wait at least 10–15 seconds for the operating program to register the change. Remember, it only takes a reading every 5 seconds, so it needs the time to register the movement of the antenna. Carry on turning the antenna like this and if any large peaks are seen click on the top of the peak and jot down the frequency, power, and strength. It is also a good idea to mark the floor in some way to denote the direction of the antenna at that particular moment. Something as simple as a chalk mark will do. After completing one whole revolution, it should be possible to have an idea where to position the antenna to get the maximum gain from the signals.

If no signals are detected, backtrack and work through each step again and double check that nothing has been overlooked. If there is still no signal, either find a quieter site with less interference to use the antenna, or build a larger antenna, or both.

Assuming there is a signal, place the antenna to get the maximum gain from the signal. The thinner and taller the peak, the better. The old saying "quality is better than quantity" is true here, although the SuperSID monitor is capable of monitoring six individual signals simultaneously, it is better to have two good strength signals

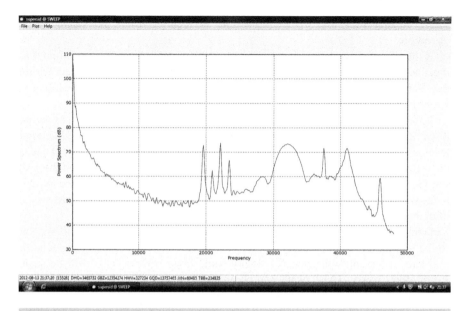

Fig. 5.11 Screen shot to demonstrate useable stations. The four peaks just *left of center*

rather than four mediocre ones. Even if only one good signal is detected the SuperSID monitor will still work. Make a more permanent mark on the ground to save going through the turning process again, or worse, to protect from some over-curious person moving the antenna in order to admire the craftsmanship of its construction.

The next part of the setting up process will take 2 or 3 days to perform. In earlier chapters we discussed the ionosphere and how during the day it becomes ionized and this is how the SuperSID monitor works by monitoring the changes to the signals reflected back from it. The ionization is at its greatest during the hours either side local midday. It is therefore a good idea to take note of the peaks that are visible around midday. All that needs to be done is to keep a note pad at the side of the computer, and each day, say over the weekend, look at the screen at midday and observe which peaks are the tallest. Click on these peaks and jot down the frequency, power and signal strength. If it is possible to do this for more than two consecutive days, maybe over the course of a week, a pattern will emerge showing which are the strongest signals. See the screenshot below for an idea of what to expect. (Although again this is far from an ideal graph, it still works well and flares as small as C1.5 have been detected on a very good day) (Fig. 5.11).

Within the SuperSID manual or via the Stanford website a list can be obtained of VLF stations and their operating frequencies and location (latitude and longitude). Each station is denoted by a three letter ID for example GBZ this is the ID for the Authorn station used by the United Kingdom, and its operating frequency is 19.6 kilohertz. Looking at the data that has been collected over the last few days

5.6 Connecting Everything Together

while running the SuperSID, a comparison of the frequencies can be checked against the list of VLF station frequencies. These are the stations that the equipment can receive. It may be surprising to see the distance that the signal has travelled from the transmitter to the equipment. The next thing to do is to tell the operating program of the SuperSID which frequencies to monitor. This is done by accessing the parameters information again. Once the parameter file has been opened, scroll down until Station 1 is seen on the list. This can be changed in the same way that the site name and location were changed earlier.

Below the heading of Station 1 is a list of three items: (1) call sign, (2) color and (3) frequency. At the side of the call sign, enter the three letter station ID, for example GBZ. The color can be left alone. "r" for example stands for the color red, and this means that this station will show up on the finished graph as a red line. This is useful when comparing more than one station at a time. If adverse to the color red for whatever reason, maybe due to color blindness, the color setting can be changed.

Beside the frequency enter the station's frequency. This must be typed in full, for example as "196000" and not "19.6". Do this for each of the stations on the list, then close the parameter file.

To confirm that the stations that have been programmed into the SuperSID's operating program are in fact stations, and not just an annoying bit of persistent interference, there is one further test to perform. Run the SuperSID monitor for at least 24 hours to get a day's worth of data. When this has been done this data can be accessed, this needs to be done while the SuperSID program is running. To access the data click on the plot tab in the top left hand corner and the list of data for each of the stations that has been selected will be there. These can be recognized by the three letters of the stations ID. Clicking on the required station and click open, a graph will be produced. Look for sunrise and sunset changes within the graph. See the image (Fig. 5.12).

If a graph similar shape to the image above is produced, this means a station has been found. Remember that it doesn't matter what the graph looks like at night, when the ionization in the ionosphere is lost. We are only interested in what happens between sunrise and sunset, as this is where any activity from the Sun will be found. Other stations can be viewed at the same time by clicking on the plot tab again and click on another station. Each graph will be shown in a different color. It is a good idea to view the stations individually until accustomed to recognizing what to look for.

If unlucky enough not to have found a station, but some source of constant interference instead, do not worry. Try and eliminate that particular source of interference by turning the antenna until the interference has gone, or has at least reduced, as long as it doesn't affect other stations. If this is not possible then it maybe a case of living with the interference. The frequency of the interference should be removed from the list of stations from within the parameter file and the program will no longer save this data.

Once happy that the equipment is functioning correctly it is a good idea to permanently position the antenna in the preferred orientation, or at least make some

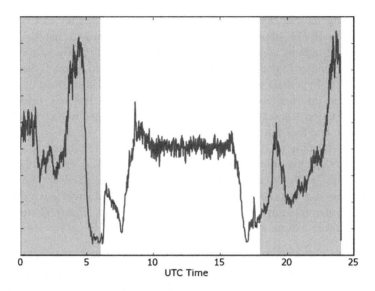

Fig. 5.12 Super-SID data showing sunrise and sunset changes

permanent marks somewhere on the ground or floor so that if any maintenance work is carried out on the antenna it can be placed back in the correct orientation once again. The SuperSID monitor's input signal can also be quite interesting to listen to through the speakers of the computer, especially if there is a thunder storm in the distance, as it picks up the lightning discharges very well. At other times it is best to mute the volume, which can sometimes sound like an insect trapped in a baked bean can.

5.7 Interpreting the Data

Now that the SuperSID monitor's operating software has been programmed and each station(s) are producing a sunrise and a sunset effect it is now time to start looking for possible solar flares. All that is needed now is a little cooperation from the Sun itself, waiting for the Sun to send out a flare of suitable size in the right direction (that of the Earth), and more importantly at the right time of day. It can be quite frustrating when waiting for a suitable flare, as they seem to have a habit of happening just before local sunrise, or just after local sunset. It's a great feeling when all the waiting pays off and a flare is recorded. In the mean time, while waiting for a flare, it is a good idea to check the graphs each day and familiarize oneself with any unusual spikes and peaks that may be found on the graphs. There are a number of different things that can affect the ionosphere and produce spikes or peaks on the graphs, including lightning, interference, or work being carried out on the transmitter itself. Lightning usually shows itself as a straight vertical line on the graph.

5.7 Interpreting the Data

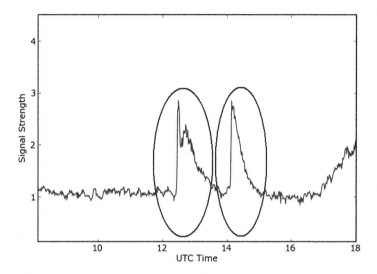

Fig. 5.13 Image showing two *solar flares circled*

Interference, this can take many different forms, and it can be of short duration or something more constant.

There is a characteristic shape to solar flares, they have a straight vertical line with a shallow angled line leading to background levels, they look like a shark's dorsal fin. Once used to recognizing them they will be easy to spot.

Looking at Fig. 5.13 it can be seen there are two flares circled. The rise in the graph that starts at about 1700 hours is the Sun setting and the ionization in the ionosphere being lost. On the above graph can be seen two flares rising vertically from the graph. Although it is perfectly possible to have the same flares going down from the line. Just by sheer coincidence two of the stations that are monitored, are GBZ at 19.6 kilohertz and the GQD at 22.1 kilohertz, have the opposite effect to each other on the graph, one is a SID and the other is a SES. Looking at the image Fig. 5.14 this effect can be seen.

The large flare on the left-hand side of Fig. 5.14 was an X3 class flare; as can be seen it is quite unmistakable. This flare left the ionosphere quite heavily ionized for some time. The flare on the right-hand side was a C2 class flare. Any suspected flares within the SuperSID, can be checked at the 3-day X-ray graph at http://www.swpc.noaa.gov/rt_plots/xray_5m.html. This graph is updated every few minutes. Please see the Fig. 5.15 below. The two flares that are circled are the same flares shown on the above image.

Looking at the graph, the X class and C class flares are both circled. There is a small trace on the SuperSID graph hinting at the next C class flare.

If the NOAA X-ray graph is consulted on a daily basis, and then checked against the SuperSID graphs, it will soon show how sensitive the equipment is, by trying to find the smallest flare the equipment can detect.

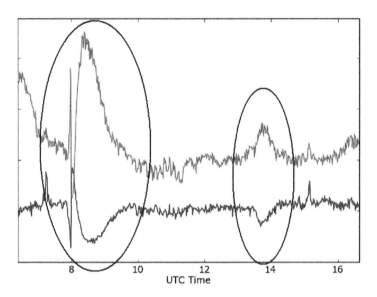

Fig. 5.14 Image showing opposite responses to the signals received by solar flares

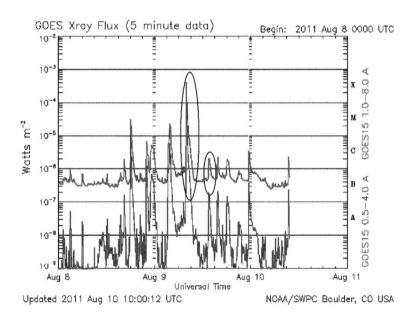

Fig. 5.15 Image from the NOAA X-ray flux web site

5.8 X-ray Classification

Solar flares are massive explosions, equivalent to many millions of tons of explosives being detonated within the solar corona. These are caused when the magnetic force lines become twisted and distorted. For example, two sunspots with opposite magnetic polarity may have a magnetic force line coming out of one sunspot and entering through the other sunspot, and because the Sun is gaseous and not solid it rotates at different rates at different latitudes. This has the effect of winding up these force lines until something has to give. It is like overstretching an elastic band, the force lines snap and a huge amount of energy is released. This is most apparent during the solar maximum when the magnetic poles of the Sun are midway between flips. The rotation of the Sun has the effect of winding up the magnetic field lines like the water coming out of a rotating lawn sprinkler.

X-ray flares are classified into five categories, A, B, C, M and X, based on increasing magnitude and intensity. Each of these categories, apart from the "X" class, which has no upper limit are further sub-divided into nine different numerical ranges, i.e. A1–A9, B1–B9, C1–C9, M1–M9, and then X1–X<. The "A" class flares are the smallest and happen the most frequently. These flares are too small to be detected by the SuperSID monitor. The class "B" flares are also too small to be detected by the SuperSID monitor. The next are the "C" class flares, and these are the start of the SuperSID monitors range. The smallest flares detectable on a good day are the C1, but more often the smallest will be the C2 flare. These smaller intensity flares can be a little difficult to spot to start with, but as the intensity increases to C5 they can get quite large. If the antenna is used at a noisy site it may be harder to detect the smaller flares.

The "M" class flares are less frequent then the three previous classes and are quite unmistakable when seen on a SuperSID graph. Warnings are issued when an M5 flare is detected, as this has the power to cause problems to satellites with sensitive electronics, communication problems and also to trigger auroras to happen near the poles of the Earth. The X-class flares are less frequent than all the other classes of flares and are the biggest and most powerful ones. These are totally distinctive on a graph, and sometimes if a really large one hits the Earth's atmosphere it can overwhelm the SuperSID monitor. This produces a peak with a flat top on the graph. These can cause the sort of damage that was discussed earlier in the section about space weather, including disruption of cell phone networks and GPS satellites, power surges within high voltage electrical power lines, etc. They also make it possible to see an auroral display at lower latitudes, even at the equator. The largest flare to be recorded to date was in the region of an X28! These flares can be lethal to astronauts, as mentioned earlier.

5.9 Resources for More Information

The SuperSID monitors are sold in collaboration with SARA, the Society of Amateur Radio Astronomers, and The Stanford Solar Center. If purchasing a SuperSID monitor for use as an educational tool in a school or university, rather than for private use, it maybe possible to apply for concessions to obtain the monitor at a reduced cost. This will need to be checked, and the details of the terms and conditions would need to be examined to see if this concession is still being offered and if one qualifies for this concession. By visiting either website a link can be found to find out how to order a SuperSID monitor. A simple online order form will need to be filled in. The usual delivery information will be needed, plus a site name for the monitor will need to be thought of. This can be anything as long as it has between 3 and 6 letters. This will then be used with the monitor's personal ID number to process the information if choosing to upload SuperSID site data back to the Stanford server. In some cases, depending on what units are available for shipping, it may be a case of leaving an email address so someone can get back in touch.

There is an online help forum where other SuperSID users from all around the world talk about their equipment and swap ideas, discuss the best antenna designs, talk about their observations and ask others, how their SuperSID equipment has picked up a particular solar event. Just about anything concerning the SuperSID monitor is discussed in this friendly forum, and questions can be posted on the forum pages and within a short time quite a few helpful answers, with possible solutions will be received.

Other useful websites:

The Stanford Solar Center website is: http:/solar-center.stanford.edu/solar-weather. To find out about the online forum and everything to do with space weather. There are lists of useful books and articles about the Sun, and links to other interesting websites. Examples of SuperSID antennas and a list of VLF stations can be found.

http://spaceweather.com
http://www.exploration.edu/spaceweather
http://www.window.ucar.edu/spaceweather
http://www.solarstormsorg/
http://sohowww.nascom.nasa.gov/

There is a short podcast about the SuperSID monitor that can be found at www.365daysofastronomy.org The date of the podcast is 21st April 2010. The podcast was written and recorded by Jim Stratigos, and he's made a cracking job of condensing the subject into a 15 minute podcast. He discusses matters with a member of the SuperSID team including about how the SuperSID monitor came into

5.9 Resources for More Information

being, and how students have found it a great introduction to the subject of studying the Sun. The podcast is a great source of information and well worth the effort of tracking it down.

Don't forget the SARA website at: http:/radio-astronomy.org This website has plenty of useful information about this and other projects. It is possible to download past copies of their electronic magazine from here, and visitors can read about what other radio astronomers around the world are doing.

Chapter 6

The NASA INSPIRE Project

6.1 What Is the INSPIRE Project?

The "Interactive NASA Space Physics Ionosphere Radio Experiments" project, or INSPIRE for short, was created in the early 1990s. An experiment was carried out using an orbiting space shuttle to transmit VLF signals towards the Earth, in order to see if these signals could negotiate the Earth's atmosphere, and more importantly the ionosphere, to determine if the signal could be detected from the Earth's surface. NASA distributed a large number of INSPIRE receivers to schools all over the United States. This was done to cover a large area in order to give the experiment the best chance of success, and this was the start of the INSPIRE project.

The INSPIRE receiver is designed to receive frequencies from 0 hertz to 10 kilohertz. This frequency range covers the designated frequency bands of VLF (very low frequency), ULF (ultra low frequency), SLF (super low frequency), ELF (extremely low frequency), and into the Sub-Hz (sub hertz) range. At these frequencies all manner of strange sounds will be heard both of natural and manmade sources.

The receiver is supplied in kit form and needs to be assembled by the user. The kit contains approximately 60 electronic components such as resistors, capacities, diodes, etc., plus all the sundries such as a box, battery housing and assorted nuts and bolts. The 18-page A5 sized assembly instruction manual that is supplied with the kit is more than adequate to build the receiver, although the supplied manual had the odd image missing. But this wasn't a problem as a complete A4-sized assembly instruction manual can be downloaded and printed out from the website http://theinspireproject.org. If unsure about one's ability to build the receiver it is a good idea to download the manual and study it first before ordering the kit.

Fig. 6.1 The INSPIRE receiver as it arrives after unpacking

The A4 manual being bigger is also handy to have when building the receiver, as the print is easier to read.

As mentioned earlier, if the art of soldering and component identification can be mastered there should be no problems in building this receiver. The finished receiver doesn't require "tuning" unlike the Radio Jove receiver and the only test that needs to be performed is to switch it on and listen. This kit is an excellent starting point on the road to building radio receivers, as there are only a small number of parts to contend with and the instructions are very clear and easy to understand. The project, when built, is a highly portable receiver that can be taken anywhere that is suitable for receiving VLF emissions. It is battery operated so there's no need for an external power supply, although there is a facility to plug in an external power supply if the need arises, but a good quality alkaline battery seems to last quite a while.

The kit can be purchased from the INSPIRE project website at http://theinspireproject.org, or links can be found from the SARA website at www.radio-astronomy.org.

An antenna and a ground stake will also need to be purchased or fabricated in order for the receiver to function, but this will be covered later in the chapter. A pair of headphones and a 9 volt pp3 battery to power the receiver will be needed too. After ordering the kit arrives in about 2 weeks. Figure 6.1 below shows the kit after unpacking.

Note each plastic bag is numbered. Each bag contains all the same components, such as resistors and capacitors, and just need to be sorted out to identify the value of each component.

6.2 A Guide to Building an INSPIRE Receiver

Before starting to build the INSPIRE receiver ensure plenty of time is available, and try to work where there is little chance of being disturbed. Building the receiver isn't a race; remember the old saying "check twice solder once" as this provides a good guide. The first thing to do is check that everything is present and correct. The assembly instructions have a list of the kit's contents, as follows:- A black plastic enclosure (Box), aluminum face plate, and circuit board. It is very important to handle the circuit board by its edges like handling a DVD or CD, to prevent the oils present in the fingers from coming into contact with the connections. This oil can be slightly acidic and can cause the solder not to adhere correctly to the connection. This causes what is known as a dry joint. Notice also that two components have already been fitted to the circuit board. This has been done to save any confusion, at the building stage.

Next check the bag that contains all the sundries, items such as switches, knobs, antenna terminals and other miscellaneous items. Before starting on the other bags which contain the electrical components, try and get hold of some small plain stickers; something about the size of a price tag will be fine. These can be purchased from stationers or other office suppliers. Mark the stickers, "R1" the second "R2" and so on, until there are 28 of them. Now open the bag of resistors and lay out the contents, pick up the first resistor (it doesn't matter which value it is). Using the color code, either the one within the instruction manual or the one within this book, make a positive identification of the resistor's value. Once this has been done, check this value against the list within the instruction manual, and at the side of the value there will be a number from R1-R28. Stick the appropriate numbered sticker to the wire of the resistor in such a way that it can easily be identified. Take extra care when identifying R3 and R4, R3 has a value of 2.2 Meg ohm and R4 has a value of 22 Meg ohm so these can be easily mixed up. Carry on until all the resistors have been identified.

Then start on the capacitors, but this time marking the small stickers "C1", "C2" and so on. Some components, like electrolytic capacitors, are polarity sensitive and MUST be fitted in the correct orientation; all the parts that are polarity sensitive are indicated within the assembly manual. Do this for all the other items until every one of the components has been positively identified.

This advice will be repeated when describing how to build the Radio Jove receiver. No apology is made for this, as this method may sound time-consuming but it has proved to work without fail over and over again, and will save time in the long run. Next take note of the positioning of the components on the circuit board. Unlike the Radio Jove receiver, where all the components fit on the same side of the circuit board, the INSPIRE circuit board uses both sides. The resistors, capacitors and other electronic components fit on one side, and the variable resistors, switches and LEDs fit on the other side of the circuit board. The best way to avoid this mix up is to place all of the parts that fit on the front of the circuit board like the switches, LEDs, etc., into a separate bag. Thus remembering to turn the circuit board over before fitting these components. This may sound like stating the obvious, but they can be fitted incorrectly.

Some components, like resistors and capacitors, will need to have their connections bending in order to fit them into the circuit board. This isn't as simple as it sounds. Resistors, and especially capacitors, can be easily damaged. If their leads are bent too close to the ceramic body of the resistor or capacitor, this can cause the component to fail. This may not happen straight away but could occur over a period of time, and these failed components can take some tracking down in order to replace them. A good tip to remember when bending the connecting wires is to leave approximately 3 millimeter (0.125 inch) either side of the components body. This can be easily done by using a pair of needle nose pliers. Holding the connecting wire with the needle nose pliers as close as possible to the component's body, then bend the connecting wire to the correct angle with the fingers, the pliers will help support the components ceramic body and stop it from being damaged. Once one side has been bent to the correct angle, if the component is then placed in its position in one of the pre-drilled holes on the circuit board, the other side can then be marked with a marker pen so knowing where to bend the other connecting wire. Once familiar with the layout of the circuit board and where each component needs to be fitted it is time to start soldering each component in place.

The best components to start with are the resistors, as these are less bothered by the heat of soldering, and also resistors are not polarity sensitive and so can be fitted in any orientation. Find the position of the first resistor "R1", look through the resistors and find the matching resistor marked with the sticker "R1", then bend the connecting wires, and solder it in place. Using a good sharp pair of wire cutters, cut off the excess wire after soldering. It's a good idea to have a small container handy to drop these excess bits of wire into, as they can be a problem if they are dropped on the floor and stood on. This can be a real problem to those with pets, as a sharp bit of wire sticking into a pet's foot can be very painful.

When all the resistors have been soldered in place, solder the two IC sockets onto the circuit board. Keep the ICs themselves in a safe place until later, as ICs can be easily damaged and should be one of the last components to be fitted onto the circuit board. It is very important that the sockets are soldered in the correct orientation so that the ICs can be correctly fitted later in the process. Look at the socket to see the small notch in one side. This notch denotes the top of the IC that will be fitted later. Look closely at the circuit board to see the corresponding mark where the IC socket is to be fitted. These marks should be lined up together to make fitting the ICs easier later.

The next group of components to solder in place are the capacitors. This should be done in the same way as the resistors, starting with "C1" and steadily working one by one through them all. Some of the capacitors are polarity sensitive, and these MUST be fitted in the correct orientation for them to work correctly.

Be very careful with the ceramic bodied capacitors as these are fragile and can be easily damaged when bending the connecting wires to allow them to be fitted on to the circuit board. The electrolytic capacitors are the ones that have a small metal casing that looks like a small can. When bending the connecting wires on this type of

6.2 A Guide to Building an INSPIRE Receiver

capacitor try not to let the connecting wires touch the metal casing, as this could cause a short circuit. The metal casing should be insulated, but it is just not worth the risk.

Next fit the diodes, but remembering from the section on component identification, a diode is a semi-conductor that allows the flow of electrons in one direction only, therefore these are most definitely polarity sensitive.

Looking closely at the body of a diode it will be noticed that the manufacturer has marked the diode with a solid color band all the way round one end of the diode's body (the color of the band is not important). Holding the diode horizontally in the right hand with the color band also on the right means the current flow would be from left to right. The circuit board is marked accordingly, so all that needs to be done is match up the color bands with those marked on the circuit board. There are zener diodes also included within the kit and these must not be mixed up with the other diodes. Within the construction manual there are images showing the different types of diodes and it is very easy to identify which is which.

Now the two inductors "L1" and "L2" need to be fitted. These are not polarity sensitive and their positions are clearly marked on the circuit board. To fit these very little bending of the connecting wires are needed to fit them in place.

The battery holder needs to be fitted now. This is just a matter of applying four nuts and bolts to hold the assembly to the circuit board, and soldering the battery leads to the circuit board to apply power to the circuit.

The positive and negative connections are clearly marked on the board. Please note that there are four nylon washers that need to be fitted between the spacers and the nuts. These are important for later, when the aluminum face plate is fitted. Working from the other side of the circuit board, it is time to fit the switches, etc. that were placed in a separate bag earlier to avoid confusion. Start by fitting the two variable resistors (potentiometers) "R7" and "R26". The circuit board is very clearly marked for these next few components. The connections for "R7" and "R26" should be pushed all the way through the pre-drilled holes in the circuit board until the back of the casing is resting on the circuit board. They will only fit one way so there is no worry about fitting them incorrectly.

Sometimes the connections are overly long, and a little of the connection can be trimmed off before soldering them in place from the reverse side. The switches should now be fitted. There are five of them with six connections each. Working one switch at a time, push each switch connection through its pre-drilled holes in the circuit board. Push the switches firmly until the casing of the switch is resting on the circuit board, and solder them in position. The same applies to the switches as "R7" and "R26", they will only fit one way. Once all the switches have been fitted the two LED's should be soldered in place. LED's, being diodes that just happen to glow when electricity is passed through them, are polarity sensitive and must be fitted in the correct orientation. Looking closely at an LED it will be clear that at the base of the bulb shape there will be a rim or collar running around the base. One side of the rim will have a flat section that denotes the negative connecting wire. This is shown on the circuit board, so just orientate the LED so the two flats, the one on the LED and the one on the circuit board, line up.

Next the ICs can be fitted into their sockets. This must be done with great care as they are very easily damaged. The first thing to do is sort out which IC is going into which socket, since there is only just the two this should be quite straightforward. Next, orient the IC so it can be fitted the correct way round; this was why it was so important to fit the IC sockets correctly earlier. Looking at the IC there is a mark near one end. This mark can be a notch in the center of one of the ends or in some cases a colored spot is used to denote pin number one. If the IC is turned so this mark is at the top left, or in the case where there is a notch in one end this notch is at the top, the IC is now in the correct orientation. Looking at the IC socket, soldered on to the circuit board, if the circuit board is then turned until the notch in the IC socket is at the top or 12 o'clock position, then pin number one is now top left. If the IC itself, is then turned so its notch is in line with the notch on the IC socket then the IC will be in the correct orientation and can be fitted into its socket. This is covered in greater detail in Chap. 4 "component identification" where an image will be found explaining the correct orientation and pin identification.

Once sure of the orientation of the IC to its socket, it is now time to fit it in place. Check before fitting that none of the connecting pins are bent. If they are, carefully straighten them with a pair of needle nosed pliers. Offer the IC to its socket and take note if the connecting pins are spread too far apart for them to fit in the socket. If they are, carefully holding the IC between finger and thumb apply slight pressure to one side in order to bend the connections in a little. This is best done by using the surface of a desk or a table in order to apply equal pressure to each connection and hopefully move them all by the same amount.

Offer the IC to its socket again and repeat this operation until the correct spacing is reached. Once the spacing is right, the IC can be carefully fitted into its socket. Fitting an IC for the first time can be a heart in the mouth moment, but by just following a few simple precautions it can be easily done. Care must be taken to make sure that only one pin enters the connection within the socket. This may seem like stating the obvious, but it can happen and this is why it is important that the connections on the IC are not bent. It is also useful to work the IC into its socket a little at a time in a slight rocking motion, rather than applying all the pressure to the center of the IC itself. Doing this the IC is less inclined to split open.

The next job to do is fit a number of short lengths of wire to the circuit board to allow connections to be made to the data and audio outputs on the aluminum face plate.

There are connections for an external power supply, the antenna and ground stake. These connections are sometimes referred to by electricians as "jumpers". All the wire needed to make these is included within the kit, and are color-coded to make the job of assembling them easier. The length at which to cut each wire is clearly shown within the assembly instructions. Stranded wire is used, so cut approximately 3 millimeter (0.125 inch) of insulation off each end of the wire being careful not to damage the strands beneath. Then, twist the strands together at each end and apply a little solder to stop the strands untwisting themselves. This also aids in soldering the ends of the wire to the circuit board and the other outputs.

Once each wire has been soldered on to the circuit board, it is time to fit the output plugs, etc. to the aluminum face plate. This is a simple job of just passing

6.2 A Guide to Building an INSPIRE Receiver

Fig. 6.2 The finished circuit board of the INSPIRE receiver

the plug through a pre-drilled hole in the face plate and securing them with a locking washer and a nut. These nuts only need a "nip" with a pair of pliers to hold them in place. Please note that within the assembly manual some of these output plugs are fitted at a slight angle, as opposed to being straight; this is to allow them to be fitted into the box later.

Now the input and output sockets have been fitted, the face plate can be fitted to the front of the circuit board. This is done by removing the nuts and locking washers on the two variable resistors and placing the face plate in position and refitting the locking washers tightening the two nuts, and fitting two screws into the spaces of the battery compartment fitted earlier. Now solder the last few connections to join the input and output sockets to the circuit board, also solder the connection for the antenna and ground stake which will be fitted later.

The final item to be fitted is the knobs on the front of the face plate that operate the variable resistors to alter the "Data level" and the "Audio level". It's a good idea before fitting these knobs to the shafts of the variable resistors to turn both shafts as far as they will go counter clockwise, and then fit the knobs. This will position both reference markings on the knobs at the same point on the scale printed on the face plate. After this has been done it should look like the Fig. 6.2 below.

The next thing to do is to fit the completed circuit board and face plate into the enclosure. This is just a matter of applying four screws, one in each corner of the face plate, to hold both the circuit board and face plate to the enclosure. The image below shows the completed INSPIRE receiver (Fig. 6.3).

As can be seen from the above image the completed receiver is a neat little project that is highly portable.

Fig. 6.3 The completed INSPIRE receiver

It is now time to fabricate or purchase an antenna in order for the receiver to pick up a signal. Looking at the image of the finished receiver it will be noticed that in the top left hand side there a BNC fitting for use with an antenna with a BNC fitting. This plug must NOT be used to mount an antenna, as if an antenna is fitted to this plug the receiver will not work correctly. The INSPIRE receiver is designed to work with an antenna with a length of between 1 and 3 meters (39–117 inches). This can be anything from a stout piece of wire to something more elaborate, like a telescopic whip antenna. For the purpose of testing the receiver a 1 meter (39 inch) long piece of stout wire will be fine. The wire can be fitted directly to the INSPIRE receiver by a screw connection on the top right hand side of the receiver's face plate. It is a good idea to fold over the last 13 millimeters (0.5 inch) of the other end of the wire and apply two or three turns of electrical tape to cover any sharp ends in order to stop any potential eye injuries if someone should walk into the end.

A ground stake is needed because this type of receiver needs a reference point and the best reference point for this receiver is the planet Earth. But for the purpose of testing it is possible to use oneself, by touching the aluminum face plate of the receiver with a finger. This will do for testing the receiver, but the receiver will work more effectively if fitted with a good ground stake. A ground stake can be fabricated from almost anything that is made of metal and can be driven into the ground. Some ideas: a long nail 150 millimeters (6 inch) or longer, an old metal tent peg, even a corkscrew stake of the sort used to tether dogs. The ground stake must be able to have a connection fitted to it, although this need not be permanently fixed in place.

6.2 A Guide to Building an INSPIRE Receiver

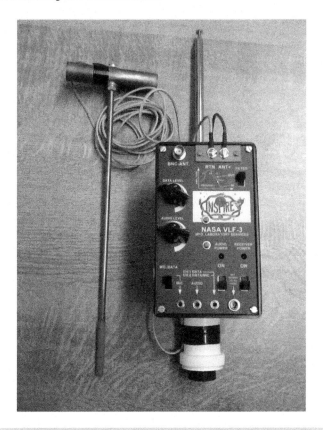

Fig. 6.4 The INSPIRE receiver showing antenna and copper ground stake on the left

The connection can be made using a crocodile clip on the end of a piece of wire which can be clipped onto the ground stake. It is also possible to use a vehicle's bodywork as a ground point, as long as the engine is NOT running. Looking at the Fig. 6.4 the ground stake is on the left hand side. It is approximately 300 millimeters (12 inches) in length and made of copper a handle was fashioned from a short length of 22 millimeters (0.886 inch) copper pipe. The tip of the stake has had a point filed on to it to help when the stake is pushed into the ground. The top of an old felt tip pen stops the point becoming a danger when not in the ground.

Note the antenna at the top of the receiver. This is a ten section whip antenna purchased quite cheaply from a local electronic supplier. It measures approximately 1.52 meters (60 inches) in length and works very well with this receiver. Alternatively, an antenna removed from an old radio could be used. The antenna could be made as mentioned above from a length of copper wire, but a whip antenna takes up far less room, being telescopic, and therefore is far easier to transport. The antenna is fitted onto a short length of 32 millimeters (1.25 inch) plastic waste pipe, the type used to take waste water from a washing sink to the drain. This plastic waste pipe is cheap, non-corroding and doesn't need painting. It can easily be purchased from almost any DIY/hardware store. It is a good idea to buy a couple of end stops at the same time as these can be fitted to each end of the pipe and can easily be drilled to accept the screw fittings for the antenna. The other end of the waste pipe here has had a ¼ inch Whitworth screw fixing fitted so that the whole unit can be mounted onto a tripod. The plastic pipe is fitted to the back of the receiver using two spring clips similar to the ones that may be seen in a workshop holding hammers and other tools to a board.

Figure 6.4 is for guidance only. Others will have their own ideas for the design. By visiting the INSPIRE website different designs can be found. One interesting design is the "Walking Stick". This design has the receiver antenna and ground stake all mounted onto a piece of 50 × 50 millimeter (2 × 2 inch) timber and all that needs to be done is place it into the ground and it is ready to use. This receiver can also be used with a wire antenna that has been suspended between two supports; a simple washing line with a wire core could even be used and suspended between two trees. The differences in designs are only limited by the imagination.

Due to the nature of VLF signals and the problems from interference it may be necessary to travel to a "radio quiet" observing site. The above compact design came into being because of this problem of interference. Everything, including the tripod, is designed to be fitted into a backpack and easily carried to a quiet observing site and quickly assembled once there. This compact design means that it is also ideal for taking on camping holidays or day trips to the countryside which are less bothered by interference. After a makeshift antenna and ground stake have been fabricated, it's time to test the finished receiver.

A pair of headphones will also be needed along with a battery, to power the receiver. Before switching on the receiver, turn the audio level knob on the front panel as far counter clockwise as possible. This will save the ears from the possibility of a very loud noise. Next, plug the headphones into the audio output socket; the second from the left. Then switch on the receiver power switch. On the right hand side an LED will light up but no sound should be heard at the moment. Now, if the audio power switch is operated the second LED should light up, and it should be possible to hear a noise coming from the headphones. Slowly turn up the audio volume on the receiver until an awful buzzing noise is heard. This buzzing noise is coming from the main electrical wiring within the home and is called "mains hum". The frequency of this humming noise is the same frequency at which the main AC electrical power is supplied to the home. This frequency can change from country to country. For example, AC current alternates of 60 hertz are used in the United States and 50 hertz in the United Kingdom. So the pitch at which mains hum will be heard is slightly higher in the United States.

Once the mains hum as been heard, if the receiver is then moved closer to a power point, the humming gets louder without altering the volume control. Holding the receiver at the side of other electrical devices such as television sets and lights it will be possible to hear other noises mixed in with the mains hum. Low voltage desk lamps and other devices that contain transformers seem to be particularly

6.3 Manmade VLF Radio Emissions and How to Choose an Observing Site   119

noisy. If these noises are heard it means that the receiver has been built correctly and is working fine. This is the first test completed.

The next test is to try and record the signal from the receiver. A simple way to check this operation is to use a recording machine such as a small cassette tape recorder plugged into the data output via a 3.5 millimeter stereo jack plug. The volume level when recording from the INSPIRE receiver is done via a separate volume control on the receiver, this is the data level control which is situated directly above the audio volume control. This allows the recording level to the recorder and the headphones to be controlled separately. To start recording first switch on the INSPIRE receiver and press the "record" button on the recording machine. Allow it to record for a minute or so, then playback the recording. If a noise has been recorded, this will be enough to test the data output socket for now. This proves that all the operations of the INSPIRE receiver are working correctly. A later section within this chapter will cover recording these types of signals in more detail.

Notice that the receiver is fitted with a built-in filter operated by a switch on the top right of the receiver. The job of this filter is to help reduce the interference problem by narrowing the window of frequencies that the INSPIRE receiver can receive. If it is found necessary to use this filter by all means use it, but experience has shown this filter is best left switched off, as the action of the filter reduces the operation of the receiver by too great a margin. It is best to treat this filter like a light pollution filter used on an optical telescope – it is alright as a compromise, but it doesn't beat a truly dark sky. Therefore it is better to try and find a quieter observing site away from interference and radio noise than to use this filter. This concludes the building and testing of the INSPIRE receiver. The question that arises now is: what type of signals should we expect to hear from the receiver?

6.3 Manmade VLF Radio Emissions and How to Choose an Observing Site

The biggest problem in relation to VLF reception is that of interference, as this can come from so many different sources. The worst source by far is that of mains hum, be it 50 hertz or 60 hertz. As nearly every home has some sort of mains/grid power it can be a little difficult to escape it. This frequency seems to be very penetrating, and if it could be seen, it could be likened to that of light pollution given off from a low pressure sodium street light. Casting its dirty yellow-orange glow over everything at night, making it hard to identify the true color of things such as vehicles, with the overspill turning the sky a dark orange color. Unlike the optical astronomer, who isn't bothered how their neighbor uses their electrical power as long has it isn't to flood light their garden. The VLF radio observer is interested in every electrical appliance because of the interference that can come from it.

Some electrical devices are worse than others. Devices with transformers, such as low power desk lamps, television sets, florescent lights and dimmer switches, are particularly bad, while others, like the normal incandescent light bulbs, are not. Although these types of bulbs are being replaced by the more efficient energy saving bulbs, these are the ones that take 5–10 minutes to "warm up" after they have been switched on. If living near an industrial area there will be a battle against the noise from electric motors. These motors are used for countless applications. Welding equipment is also bad for giving off interference, even if the equipment is fitted with a suppresser. High power switching gear can also be a problem. Overhead power lines can also contribute other problems. If the insulators on the pylons which carry the overhead power line are in poor order this can cause interference to be generated around the insulators themselves.

Sometimes there are signals that are superimposed within the frequency of the high voltage electricity. These signals can be instructions for smart meters to switch from a high tariff to a lower tariff and vice versa. These signals can produce intermittent problems to radio receivers. Underground power cables are better due to the fact that their interference is somewhat contained within the ground. It has been found that their ability to cause interference depends on a number of different factors, such as the type of ground that the cable is buried in and the depth at which it is below the surface. If everything has been taken into consideration, and as much distance has been placed between the VLF receiver and the nearest main power source, but the receiver is still receiving mains hum, then it is quite possible that there is an underground cable somewhere in the vicinity.

While walking along a beach using the INSPIRE receiver it was still receiving a faint mains hum at a particular place along the beach. There were no signs of human electrical activity for about 800 meters (0.5 mile) in either direction, but the faint mains hum could definitely be heard. It wasn't until after dark that the cause of this mains hum was found. A few hundred meters out to sea was a collection of rocks with a warning buoy amongst them. It was fitted with a light that flashed on and off every few seconds. During the day this warning buoy had gone unnoticed, but at night the light was clearly visible. The receiver had been picking up the buried cable to the warning buoy's light.

Mains power hum has another annoying problem that an VLF observer should be aware of, it can manifest itself at other frequencies. This is called "harmonics".

A harmonic is a frequency that is an integral multiple of the original frequency. For example, if mains frequency is 60 hertz then it is possible to receive a signal or harmonic at multiples of 60 from the original 60 hertz source, frequencies such as 120 hertz, 180 hertz and so on. If the mains frequency is 50 hertz then harmonics maybe received in multiples of 50. Vehicle ignition systems can also be a problem for the VLF radio observer. Luckily these can be more of an annoyance rather than a serious problem, because most vehicles will be only passing by. A good idea is to use the INSPIRE receiver at least once in close proximity to a vehicle so as to learn what to listen out for, as a single cylinder motorcycle or lawn motor engine can sound quite different to a multi-cylinder car engine.

Clothing is another source of interference of false signals, as was found when testing the finished INSPIRE receiver for the first time. Already aware that the receiver is sensitive enough to pick up the very low level VLF emissions produced by feet on the ground even while walking on grass, listening intently to the sounds from the receiver a strange "whooshing" sound was heard, it turned out to be the low level VLF emissions generated from one's clothes, which were rubbing together while walking. Analog wristwatches with a second hand can also be picked up by this sensitive receiver; a pulse can be heard every second as the second hand on the watch moves. Digital display watches don't suffer from this problem. Cell phones can be a problem, so keep them as far away from the receiver as possible. Although not a manmade source of interference, if a flying insect, something along the size of a bee, flies too close to the whip antenna (if this type of antenna is used), then the receiver will pick up the minute electrical signals given off by the insect's wings, and this produces a buzzing noise in the headphones.

Due to the problem of mains hum it is more than likely that it will be necessary to travel to get away from this problem. Here are a few important points to consider when choosing a suitable observing site to carry out VLF studies. Try to get at least 800 meters (850 yards) away from any form of mains/grid power cables, including overhead power lines, as this will give the best chance to hear the natural radio emissions and avoid mains hum. This distance should be considered a minimum and if it is possible to get further away it will be well worth the effort. Avoid valleys that are completely surrounded by hills and mountains in every direction. These will help shield from unwanted interference, but they will also shield the radio signals that we want to receive. Avoid trees, especially very tall ones. The odd tree here and there is unlikely to make a great deal of difference but don't try and observe from a forest or a clearing in the middle of a forest. Trees are living breathing organisms, requiring water to live and function (a lot of water in fact). Most of a tree's water is contained within the truck. The tree's vascular system means that the water circulates within the first few inches of wood under the bark. This is called the "sap" wood and is the new growth of the tree. This sap wood is the soft living part of the tree trunk, the center of the trees trunk or the "heart" wood is basically dead and is there only to add strength to the trunk to support the tree itself. This is why some old trees may look as if they are dead because their trunks are hollow, but what has happened is the center of the tree's trunk, the heart wood, has rotted away. They are still alive because the living part of the trees trunk the outer most part just under the bark is still alive and carrying water up to the branches and leaves. This means that a wide tree trunk makes a very good barrier to block the electromagnetic radiation that we want to receive. This is also the reason why standing under a tree during a thunder storm is a very bad idea.

Other structures to avoid include wind turbines. Whether it's love them or hate them, everyone seems to have an opinion on them. They do produce green electricity but only when the winds blowing, but they haven't been the mass killer of birds that some people thought they might be, although they have been proven to kill other animals, namely bats. Not by the action of the bats flying into the blades, but by the sudden drop in air pressure from one side of the blades to the other. This sudden change in air pressure causes the delicate membranes within the lungs of the bats to rupture and burst, killing them. There have been a number of articles which indicate that wind turbines can generate strongly within the ELF and VLF frequency ranges. This can be a menace to the VLF observer. The articles didn't say exactly how they generated these frequencies. Whether it is from the action of the blades on the air molecules or from the turbine windings by the action of the rotor turning within a magnetic field to generate the electricity, it could even be a combination of the two, but whatever the case, they should be avoided.

6.4 Natural Radio Emissions

Optical astronomers, either professional or amateur, can turn their telescopes to the skies and observe and study many things in the universe, from the most distant galaxy to our nearest neighbor the Moon, but there is one object a planet that we cannot turn our optical telescope to in order to study it, and that is the planet Earth. Using VLF frequencies we can observe and study the very planet we live on, and listen and record the normally invisible interaction between the forces of nature acting upon it.

There are a number of natural radio emissions that are picked up with the INSPIRE receiver. It is probably best to discuss each type and their origin in turn. All these sounds are heard together, but some are more common than others. The first VLF signal that is likely to be heard is Sferics. The word "Sferics" is not a description of the sound that will be heard, but is made from shortening the word "atmospherics", because at the time these sounds were first heard no one was really sure what they were and they had to be given a name, so the word Sferics came into being.

Sferics are of very short duration, typically lasting only a few milliseconds.

Sferics can vary in density; some days there are very little, just the odd click every now and then, while other days quite noisy and sound like the rustling of a bag of potato chips or like the sizzling noise of bacon frying in a hot pan. When there is a lot of these Sferics it can sometimes make it difficult to hear any other VLF signals.

The odd "click" or "hiss" heard can be caused by a random bit of static within the Earth's atmosphere, but the main cause of Sferics is lightning. The electrical potential between the Earth and the thunder clouds builds up until something has to give. The air between the thunder clouds and the Earth is usually a good insulator, but any insulator will conduct electricity if there is enough power to force the electrons through the medium. This is like over-stretching an elastic band until it reaches a point where it snaps, releasing a large amount of energy in a very short time. As the lightning bolt strikes there is a massive release of electrical energy, so much so that sometimes it is possible to smell the after-effects of the electrical charge in the air. For a brief time the air molecules that usually act as an insulator are overwhelmed and they conduct the lightning bolt. The thunder is generated by the air around the lightning bolt being superheated. This sudden heating of the air makes it expand very rapidly and creates the recognizable noise that is thunder.

Guglielmo Marconi's first radio transmitter worked by generating a spark, and it was the radio emissions from this spark that were received giving birth to radio.

6.4 Natural Radio Emissions

As lightning can be thought of as a spark, granted a very large spark, when this energy is released it also releases electromagnetic radiation as radio waves, and this is the origin of the radio emissions that will be received. In an earlier chapter we discussed the reflective qualities of the ionosphere and how it allows radio waves to "bounce" off it and travel many kilometers from their origin. This is how Sferics can travel a few thousand kilometers, by bouncing between the ionosphere and the Earth. If the thunder storm is close to the observing site the intensity of the Sferics will be greater, and if they become very loud it may be a sign that the storm is approaching or that it is a particularly bad storm.

The next radio signal noises that we shall discuss are "tweeks". Tweeks are a short duration VLF emission lasting in most cases a tenth of a second. The sound of tweeks is hard to describe, but once heard they are quite easy to recognize. Their sound has been likened to a single chirp from a bird, but with a more metallic sound in nature.

The origin of tweeks is the same as that for Sferics, lightning, but with one difference. In an earlier chapter we discussed the Earth's ionosphere and how it was made up of layers. These different layers are at different altitudes. Tweets are Sferics that has travelled up through the Earth's atmosphere, been reflected back towards the Earth's surface from the uppermost layers of the ionosphere. This allows the tweeks to travel tens of thousands of kilometers.

The change in sound of Sferics and tweeks is due to "dispersion". Dispersion is the separation of the radio emissions into its component frequencies. To give an optical example of the principle of dispersion, think of a prism and the effect it has on light. All the optical light wavelengths together make up white light, if white light is passed through a prism on leaving the prism the white light will have been split up into its component colors or wavelengths. This is due to the different energy levels within the different colors, and this produces the familiar rainbow spectrum. In the case of radio wavelengths, our prism is the reflective layer within the ionosphere, and its ability to reflect different wavelengths in different ways, this combined with the angle at which the radio waves hit this reflective layer helps split the radio waves into their component parts, as each different frequency will reflect slightly differently. This has the effect of spreading out the radio wavelengths like an optical wavelength spectrum.

The next and most "spooky" sounding VLF signals that can be heard are "whistlers". There are several different types of whistler, and they do sound as if someone is whistling. The first documented evidence of the existence of whistlers was way back in the late 1880s. They were heard through an un-amplified telephone earpiece of the time. This telephone ear piece was connected to a telephone line that ran for several kilometers, and the whistlers couldn't be heard all the time but only on certain occasions, which must have had the operators of the telephone lines scratching their heads as to what these sounds were and where they were coming from. What wasn't realized was the long length of the telephone line was acting as a large antenna and collecting the VLF radio signals.

The nature of a whistler and their unusual sound made speculation grow as to their origin. Could their existence be due to lost souls trapped between this mortal world and the next? This belief was not uncommonly used to explain the unexplainable and strange phenomena. There was an ancient Norwegian belief that the aurora that flickers within the Earth's atmosphere was the souls of the dead being transported to the afterlife. Others thought it could be proof of the existence of extraterrestrials and this was their way of trying to contact the Earth. Investigations took place to try and find out what caused these strange sounds and where they could originate from, but whatever these whistlers were the technology wasn't advanced enough to work it out and some of the interest in these strange sounds faded. It wasn't until the outbreak of World War One when the interest in the strange whistling noises came back into vogue.

Radio communication was still in the very early stages of development. The equipment needed to make and receive radio transmissions was large and required a lot of power to operate it. Plus, any radio transmissions would probably have been very easy to intercept by the enemy. During World War One the fighting was conducted from trenches and a suitable and reliable way was needed for the commanding officers, to communicate with the officers in other trenches. So, a network of telephone cables were run from trench to trench, and some of these trenches were large distances apart. At certain times of the day, but not every day, a strange whistling and hissing noise could be heard coming from each trench telephone ear piece. The first thought was that these strange noises was an attempt by the enemy to block or listen into the communications. Foul play by the enemy was quickly ruled out and it wasn't until the end of the war when the mystery of these strange noises would start to be understood.

As the technology had yet to be developed, the existence of the sounds relied on the testimony of those who had heard them. After World War One, tape recording machines were starting to be developed. These tape recording machines, by today's standards, would be classed as rather crude, and the sound quality wouldn't have been great, but it would have been enough to record the strange sounds, enabling them to be processed later. The sounds were processed using a basic spectrogram. A spectrogram plots frequency changes against time (spectrograms will be covered later in the chapter). Using spectrogram analysis it was found that these strange noises were whistlers and other ionospheric phenomena, and that the long telephone lines which ran between the trenches were acting as a large antenna.

It was discovered that there were different characteristics to some of the whistlers that had been recorded, this gave scientists a clue to their origin. It was found that whistlers originated from bolts of lightning on the Earth, but unlike tweeks where the radio emission gets trapped between the Earth's surface and the top layers of the ionosphere and bounce between the two, a whistler gets trapped within the Earth's magnetic field. As discussed in an earlier chapter, the Earth's magnetic field is produced by a dynamo effect caused by the rotation of the Earth and it's inner and outer cores, this magnetic field can stretch a great distance into space. It is thought that the radio emission created by the lightning bolt travels up through the Earth's atmosphere, through the ionosphere, and gets trapped within the magnetic field lines of the Earth. It then can be carried along the magnetic field line right into space before being returned back to the Earth with the magnetic field line as it re-enters the Earth at the opposite pole. Whistlers can travel great distances around the Earth, and it is not uncommon for them to travel from one hemisphere into space and return to the other side of the Earth in a different hemisphere.

Depending on how whistlers move within the Earth's magnetic field they can sound slightly different, and with the use of a spectrogram this difference can be seen visually. For greater knowledge of the different types of whistlers it is highly recommended to read "Whistlers and Related Ionospheric Phenomena" by Robert A Helliwell (Dover Books on Electrical Engineering, 2006), as this book covers whistlers in great detail.

The last VLF signal to discuss is the "chorus". The chorus is a very fitting name for this type of VLF signal, it is also fitting for the particular time of day that it can be heard. The best way to describe the sound of a VLF chorus is, first thing in the morning just as dawn is breaking, the "dawn chorus" as all the birds have started to wake up and start chirping to each other. The best time to hear this VLF chorus is also early morning. The VLF chorus is thought to be a result of charged partials contained within the solar wind interacting with the Earth's magnetosphere. These charged particles can get trapped within the magnetic field lines of the Earth's magnetic field and be funnelled down towards the Earth's poles like the aurora. This is why the chorus is best heard from mid to high latitudes in both the northern and southern hemispheres, but not very often on the equator unless solar activity is very high.

Sound Samples

In optical astronomy it is quite easy to show an image of an object or item to help demonstrate what is being explained, but unfortunately it can be quite difficult to describe in writing just what sort of sounds that can be heard while using a VLF receiver. The best way to hear a sample of the different types of emissions that will be received is to visit the INSPIRE project website. Once there a number of examples of manmade and natural VLF radio emissions can be found. Owners of MP3 players that are capable of playing WMA format sound files. It's a good idea to load the sound samples onto the MP3 player and play them from time to time to get familiar with the different types of sounds, this can also be done with the sound sample from the Radio Jove website. Doing this it is surprising how quickly it becomes to identify the different characteristics of each sound.

6.5 Radio Emissions from Beyond the Grave?

Quite a lot of mileage has been made about the existence of VLF and ELF frequencies by movie makers. This includes movies such as "Poltergeist" and the famous line from the little girl: "They're here" as the spirits communicate with her through the television set. Another example is provided by the movie "White Noise" where the star of the movie loses his wife tragically and he takes to recording the white noise from the television set in an attempt to hear her voice. There is even a warning at the end of this movie, just before the credits, saying not to try this as it can end with unpleasant things happening. The movie "Ghost Busters" even cashed in on VLF – ELF signals. In the movie they arrived at the haunted hotel and scanned the area for VLF – ELF signals. In reality the only thing they were likely to pick up is the mains hum from the electrical supply of the building and every other building in the vicinity.

The Radio Jove receiver is designed to work around 20.1 megahertz, and some shortwave radio stations can broadcast at a similar frequency. Even though the Radio Jove receiver is not designed to pick up voices like a normal radio, it is possible to hear an odd distorted word or something that sounds like a word every now and then.

The INSPIRE receiver is designed to work at such a frequency that it shouldn't contain any commercial broadcasts. "Shouldn't" being the word, because if the atmospheric conditions are just right something that sounds like a distorted word may be heard. The sound of some whistlers can be quite haunting and ghostly. At the website www.vlf.it, it can be seen that enquiries are often made about the possibility of using VLF and ELF frequencies to communicate with lost loved ones.

The only supernatural experience witnessed so far occurred as a result of a nights meteor observing. While setting up radio meteor detection equipment for a night's recording and forgetting to mute the computer's microphone in the audio input settings, leaving the microphone live all night. Unaware of this oversight, when processing the sound file there were some very strange noises recorded along with the meteor echoes. Were these strange noises from tormented souls trapped between this world and the next? The simple answer is no. After a little detective work and realizing the oversight it was found these strange sounds were in fact noises from the local wildlife and the observatory's squeaking door hinges (which have since been oiled).

For some reason however suspicion persists around VLF – ELF frequencies. Communications with extraterrestrials have even been suggested, but why extraterrestrials would chose to use these frequencies in not clear. Another suspicion and speculation is that VLF – ELF frequencies can harm living tissue by interfering with the body's natural resonance, such as brain wave activity, and the functioning of other internal organs, disrupting the workings of the body. Whether VLF – ELF frequencies can damage health is for science to decide, but so far it is unsupported.

6.6 Recording VLF

Recording the output from an INSPIRE receiver or any other radio receiver isn't as easy as it may at first seem. The average recording device, whether it is analog or digital, is designed to record music or speech and for this reason most, if not all, recording devices are fitted with an "automatic gain control" (AGC). An AGC is

6.6 Recording VLF

designed to keep the recording sound levels equal and to stop any sudden changes to recorded input levels. This operation is done automatically by a circuit contained within the recording device itself. This circuit monitors the inputted signal level and either increases or decreases the recording level in order to keep it as constant as possible, without any sudden changes in the sound levels when the recording is played back.

This function requires a sort of balancing act to maintain a constant level at all times. There will be an AGC fitted on to a vehicles radio in order to maintain a constant volume output from the radio and to stop the volume increasing or decreasing due to any variations in signal strength from the radios antenna. For example, when driving under a bridge the antenna signal may be reduced temporarily, but instead of the radio's volume being reduced the AGC will boost the level to maintain the volume at which the radio is set. Once the signal strength has returned, after driving from under the bridge, the gain control will automatically reduce the level again and stop the volume of the radio increasing.

To understand how these AGCs will affect a recording, we can use a practical recording situation as an example. Imagine that a number of people sit around a table for a meeting and a recording device is placed at one end of the table, to record the proceedings of the meeting. If someone sat in close proximately to the recording device starts speaking the machine will automatically turn down the gain of the recorder due to that person's voice being louder, because of them sitting closer to the machine. On the other hand, if someone at the other end of the table starts to speak, with that person being further away their voice will be at a lower audio level. The recording device will then automatically increase the gain so the recorded level remains the same as the first. Then when the recording is played back the sound levels should be at the same level or very nearly.

Taking the same example but this time make the recording on a machine with the AGC switched off or disabled, the following will be heard: the first person who is sat close to the recorder will appear to sound very loud or even sound as if they are shouting when they speak, while the second person, who is sat further away, will sound very faint with a high degree of background noise. If the meeting is being held in a large room there maybe even an echo or two recorded as well. It will give the impression that the second person is speaking from the bottom of a well. Therefore AGC's make for good recording of music and speech, but for the radio astronomer they are nothing but trouble.

When it comes to recording a signal, either from an INSPIRE or Radio Jove receiver, these types of signals are full of sudden increases and decreases in amplitude levels. For example a type of signal that maybe received from the Sun using the Radio Jove receiver, the one that produces a "shark fin" shape on a graph. This consists of a sudden increase in noise that slowly returns to background levels over a period of time. The AGC would just alter the input so that it would produce a constant noise, at best it may allow a slight increase in noise to be heard, but the full effect would be lost. The shorter duration signals, such as S-bursts from Jupiter using the Radio Jove receiver, and the tweeks and sferics from the INSPIRE receiver, would be the worst affected of all, because of their short duration, in some cases less than a second, the AGC would just "clip" these off the recording. The longer duration "L" bursts from Jupiter seem to be less affected, but these still suffer in the same way as described above in the example for the Sun. Therefore a way must be found of recording the input from these receivers without losing the part of the recording that is wanted.

A recording device must be found that has no AGC fitted, which is highly unlikely. It may be possible to find a recording device where the AGC can be switched off. There is another way to disable the AGC from the inside of the recorder, but this is not recommended unless one knows exactly how to do it. This can be done more easily on the older type of cassette recorders where it is possible to trace each part of the circuit board, and with a quick snip with the wire cutters in the right place can disable the AGC, but some cassette recorders have the AGC built into another circuit and any attempt to disable it will cause a total failure of the unit. The new digital recording devices, such as a dictation machine, may seem a good idea as they have excellent recording qualities and can in some cases be easily coupled to a computer for quick downloading of the sound file, but these also have an AGC fitted and so are of very little use to the radio astronomer. There are still some professional cassette recorders available from auction sites, where the AGC can be switched off, but these have the added problem of having to obtain the cassette tapes.

If choosing to try cassette tapes for recording, opt for the better quality chrome or metal tapes over the standard oxide type, nowadays it may be a case of getting hold of any that can be got. There are other problems with using cassette tapes for recording, these include the frequency response of a cassette which is designed to be suitable for audio frequencies of 20 hertz to 20 kilohertz and therefore any response either side of this will be an unknown entity. Cassette tapes are also not as sensitive, compared with their digital partners, therefore some weaker signals may be lost. One last thing to consider in relation to using cassette tapes, and sometimes the most frustrating, is that of the noise from the cassette itself. In the age of CDs and other digital recording devices the noise from the spools within the cassette tape can sometimes seem deafening, but it is possible to get used to this noise and in some cases remove some of it by using filters within processing programs. It has also been suggested that mini disc recorders are quite good at recording these signals.

If choosing to use a digital recorder, the format that the sound is recorded and stored in is an important factor. The sound file formats that seem to be the best are WAV, or WMA or other files of this type. Don't be tempted to use the MP3 sound file format, or formats of this nature, especially to save on storage space on the recorder. The way the MP3 sound file is recorded and stored makes it highly unsuitable for use in radio astronomy. The MP3 recording format works by using what is known as a compression algorithm to record the sound, but this compression algorithm works by removing some of the original sound recording, to save space on the recorder. This missing part of the sound file is not noticeable when listening to rhythmic pop music, if listening to something more along the lines of classical music the difference will be noticed straight away. When the first MP3 players came into being an attempt was made to convert Holst's "The Planets" performed by the London Philharmonic Orchestra, from cassette tape to MP3 format. After the piece had been converted it proved to be a bit of a disaster, as the MP3 recording was lacking in quality and some parts of the original recording were missing.

MP3 format recordings seem to be the poor relation to WAV and WMA recording. It is like comparing the image quality of a camera with 1 million pixels to one with 10 million pixels, the 10 million pixel camera will win hands down. MP3 recordings are alright for what they are intended for, something to listen to whilst jogging around the park. The best way by far to record these type of signals is to use a computer and the sound card that it contains. The quality of the average sound card can easily out-perform any cassette tape recorder, and in most cases outperform other digital recording devices. The sensitivity of the computer's sound card, both in its ability to detect weaker input signals and the frequency responses are far better. The background noise levels of the sound card are also far lower than other devices.

Operating programs such as Radio-SkyPipe, which will be covered in later chapters, can be programmed to cancel out the remaining noise from the sound card to give a more accurate and truer reading. Another very good point is that none of the computers that have been owned so far have been fitted with the dreaded AGC.

To summarize, VLF signals are best recorded on a computer via its sound card, but in cases where it is necessary to travel to a radio quiet site a computer may be impractical, so another recording device will be needed, and this can be either analog or digital. If an analog tape machine is used it must have the AGC switched off or disabled. Also use the best quality tapes that can be found, such as the chrome or metal type. If a digital device is to be used, make sure that the AGC can be switched off. The chance of disabling them on a digital device is rather slim. Also only use WAV or WMA sound file formats to record, as MP3 recordings are of inferior quality due to the compression algorithm element used in the MP3 recordings.

6.7 Analyzing Software and Where to Get It

There are a number of good analyzing software programs available which are ideal for analyzing VLF frequencies. The two that will be discussed here are free to download from the internet for domestic use only. Please see all terms and conditions of use from the website before downloading the software. Both programs are quite intuitive and easy to use. The first is "Spectran", the brain child of radio enthusiasts Alberto Di Bene and Vittorio De Tomasi. This program can be downloaded from www.weaksignals.com. Looking at the image below, a screen shot of the program showing some of the basic controls.

As seen from the image, the operating program consists of two main windows. The top window gives a graphic representation of the incoming signal, while the window below is a spectrogram image of the same incoming signal. A good way to think of a spectrogram is as a graphic representation of any change in frequency of the signal against time. For example, as described above, the noise from a whistler will be heard to change in pitch during the duration of the whistler. If this change

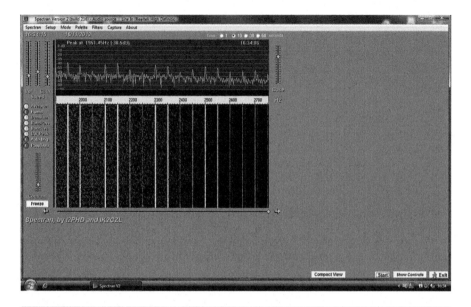

Fig. 6.5 Screen shot of Spectran

in pitch is plotted against time it will be shown as a curve on the spectrogram window.

A spectrogram is also useful for observing the random return echoes from a meteor's ionized trail. It is especially useful if the meteor breaks up in to a number of pieces, as the spectrogram will show the multiple echoes within the signal. Spectrograms can also be useful in identifying unwanted manmade signals. Manmade signals usually have a pattern to them, and this pattern can sometimes be seen within the spectrogram window, for example mains hum or the noise from a large electric motor.

This window can be used in the default setting, whereby the window shows changes in the spectrogram as a waterfall effect, i.e. from top to bottom, or it can be changed to run from left to right if preferred. The color and contrast of the display can be changed to suit personal taste. In the top left of the Fig. 6.5, the main controls are of a slider type. These control volume, speed and gain. Just below these sliders are a series of filters, one of which is very useful indeed. This filter is called the "de- humming" filter, and when this filter box has been checked it removes most of the mains hum from the signal. For this filter to work correctly the frequency of the mains hum, either 50 hertz or 60 hertz must be set first. This can be easily changed by clicking on the mode button on the top tool bar, then checking the 50 hertz or 60 hertz "De hum" box. Although this is a useful feature it still doesn't beat getting data from a radio quiet observing site.

If something of interest is seen in either of the two windows there is a freeze button that will save the screen image to a file so it can be retrieved later. By clicking the "Show controls" button, bottom right in the Fig. 6.5, it is possible to make

6.7 Analyzing Software and Where to Get It

quick changes to the sampling rates and to make other adjustments to the program even while it is running. Once the program is started, a "Record" button will appear just below the top tool bar. If this button is pressed the inputted signal will be saved, a window will pop up asking the operator to input a file name, so a file can be automatically created and this can be retrieved later. The file format is set by default to the WAV format, which is ideal. To stop the program recording it's a simple matter of pressing the "Stop" button.

The controls such as those required to record, play, and pause, are very easy to master; they are just like the controls of a DVD player. The play back feature of the program is useful for playing back recordings that have been made of VLF recordings or meteor echoes. There is also a Moon position window that, if clicked, shows the position in azimuth and elevation of the Moon. This is a great program to experiment with and the setting can be endlessly changed in a quest to find the perfect setting for the purpose. If getting into trouble with the program while experimenting, it has a "Reset" button. If this reset button is pressed the program will automatically reset itself back to default settings.

All in all Spectran is easy to use and quite straight forward to understand. Out of the two programs that are to be discussed within this section, this program seems to run better on slow computers. That feature makes it an ideal choice to use on an old laptop that is out of date and has a slow processor that can be left outside in an observatory over night as this program will run perfectly happily on it, and record data which can then be processed on a faster machine later.

The next program that is very useful for VLF studies and meteor observations is "Spectrum Lab". This is a wonderful program created by Wolfgang Buescher and can be downloaded for free, for domestic use from Buescher's website: http://wwwqsl.net/d14yhf/spectal.html. As before please see all terms and conditions of use on the website before downloading the software.

This program is being constantly refined and improved by its creator. It has a vast number of utilities, far too many to mention here as it would take a whole book of its own to do it justice. The Spectrum Lab instruction manual is well over 100 pages in length. As this program is more advanced than Spectran only the basic operation of the program is covered here. For those wishing to download this program it is highly recommended that the latest incarnation of the instruction manual is downloaded at the same time to help with some of the operations of the program. Figure 6.6 shows a screen shot of the program.

At first it will look similar to the screen shot of the program Spectran. The top window is a graphic representation of the incoming signal, and the large window below is the spectrogram window. To run this program as it should be run a reasonably fast computer with a fairly large amount of free space on the hard drive will be needed. Although this should not be such a problem nowadays with up to date machines, if certain operations are tried on a slower computer the computer is unlikely to be able to perform, and the program will usually flash up a warning saying that the computer is too slow for that operation to be carried out. If this warning is ignored and the operation is carried out anyway freezing of the program may be experienced if the computer's processor is too slow.

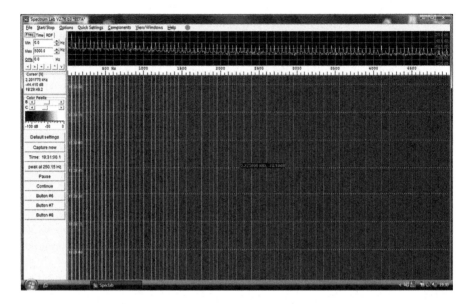

Fig. 6.6 Screen shot of Spectrum Lab

Spectrum Lab has just about everything the amateur radio astronomer could wish to have for processing their data. There are some operations that the program can perform that may never need to be used, depending on the type of signal being processed and how the operator wishes to process them. There is a very useful feature to Spectrum Lab where it is possible to change everything pertaining to the soundcard and the input mixer on the computer through a control panel window. To change or adjust any particular part of the soundcard or input mixer, just click on the item one wishes to change and another window will pop up asking the operator how they want to make the adjustment.

There are numerous options, settings, controls and filters, far more than the Spectran program. It also has the "De-humming" filter as described above. Even the color of the spectrogram screen can be changed. The program can also be used for recording the input from a radio receiver, so for instance use it to monitor and record a night's worth of radio meteor observations. It will create a sound file automatically and save it so it can be retrieved later for processing in the same sort of way as described with Spectran. It is capable of replaying almost any type of sound file. Additionally, Spectrum Lab is programmable. This can prove useful if wishing to use the program to count and record meteor echoes, although this can take some time to do and it is not as easy as it may first sound, but it is possible with practice.

There is a "Freeze" button which will automatically save a screen shot of the program, and this is useful if anything interesting within the spectrogram window is spotted. Thankfully Spectrum Lab also has a "Reset" button, as described above, that will return everything back to default settings. It must be confessed that this button has been pressed quite a number of times while trying to learn some of the operations of this program.

Of the two programs Spectrum Lab has more useful features, especially if liking to experiment with different filters, etc., while on the downside it takes longer to master. Spectran is easier to use from the start, and is perfectly adequate for the purpose of VLF processing. Perhaps the best route is to start with Spectran and then progress to Spectrum Lab after gaining some experience handling Spectran. Spectrum Lab is a really useful program, but it can take a while to become familiar with its many operations, although its basic recording and replaying of sound files is quite easy to master. One could also download both programs and see which program suits the needs of the operator the best.

6.8 References for More Information

The first place to start the hobby of VLF observing is the INSPIRE website at http://theinspireproject.org/. Once there it is possible to find all the latest news involving the INSPIRE project, as well as finding examples of different types of antennas, sound files containing sample sounds of all the sounds described above, and more. It is possible to order an INSPIRE receiver through this website.

Students can apply for grants; this will of course depend on personal circumstances and have to be investigated for oneself.

Another good source of information that can be found at the INSPIRE project's website is the INSPIRE Journal. Within the Journal can be found all the latest information, useful articles, websites, etc. Some INSPIRE users send in details and images of their receivers and their interesting antenna designs. It is highly recommended looking through some of the back issues of the Journal, if only to see how it has changed from its humble beginnings.

The INSPIRE project and INSPIRE can also be followed on Facebook, or else register one's email address through the INSPIRE website at http://theinspireproject.org/ to receive updates and other useful information via email.

Another excellent website to visit is www.vlf.it, which is dedicated to the study of very low frequency emissions. It has examples of antennas suitable for the reception of VLF emissions and lots of useful links and applications, such as the use of a processing lab for processing VLF signals. It also offers a section for questions (of the non-afterlife variety). The helpful book "Radio Nature" by Renate Romeo can also be purchased from this website.

Another useful website, this time for software, is www.weaksignals.com. This site has some useful software that may be helpful to the more advanced radio astronomer enthusiast.

Chapter 7

The NASA Radio Jove Project

7.1 About Radio Jove

A quick search on the internet will reveal a great deal of information about how to use a short wave radio receiver, to receive radio emissions from the planet Jupiter and from the Sun. Some of the information and ideas that have been posted are good, but some are quite frankly wrong. Starting with the correct equipment gives the best chance of success. To use optical astronomy again for a moment for an example, if entering the hobby by purchasing one of the "department store" telescopes with their poor optics, very shaky mounts with poor telescope controls and the ridiculous magnification claims of ×600 for a 60 millimeters (2.4 inch) aperture. They will get fed up trying to operate the telescope and will soon lose interest and move on. Whereas, the advice from a fellow astronomer would probably be to purchase a good pair of binoculars and a good star map as a first step.

Radio astronomy is no different. There are claims on the internet saying that it is possible to receive radio emissions from Jupiter with a small loop antenna of about 600 millimeters (24 inches) in diameter. This may be in the realm of possibility, but don't bet any money on it working. Also not any old short wave receiver will do, a receiver must be found where our old nemesis the automatic gain control (AGC) can be switched off, disabled, or bypassed. In many cases this cannot be done, and attempts to do so may stop the radio from working. Even if this is managed the problem of knowing whether or not the receiver was receiving the right type of signal would still be there.

Step forward the Radio Jove receiver. The Radio Jove receiver works on a narrow frequency range centered around 20.1 megahertz, and was designed specially

Fig. 7.1 The Radio Jove kit as is arrives after being unpacked

to receive the radio emissions from the planet Jupiter and from the Sun, and it does exactly that.

The Radio Jove project is an education and public outreach program involving scientists and teachers from NASA and other organizations. This program supplies a relatively cheap kit capable of receiving radio emissions from the planet Jupiter and the Sun. The kit includes parts numbering a little over 100, and the user assembles the receiver for themselves, although a pre-built and pre-tested receiver can be purchased for an extra charge. As supplied, the kit has everything needed to start on the road to running a successful solar and Jovian radio telescope. The only other items required are a 12 volt power supply (this can be a battery), a pair of headphones or external powered speaker(s), and the poles from which to hang the antenna. These poles can be metal, wood or plastic, but plastic has the habit of flexing quite a bit, especially when working at height. Figure 7.1 shows the kit as it arrives from the Radio Jove team after unpacking

The kit contains five small plastic bags which are numbered, and each one contains different components, for example resistors, capacitors, etc., along with the circuit board, power cable connections, an aluminum enclosure, antenna connections, six ceramic insulators, a coil of copper wire (from which the antenna is to be made), and a coil of RG59/U antenna cable. Two CD-ROMs are also included, accompanied by a very clear and easy to understand instruction manual. The CD-ROMs contain lots of useful information and visual aids, plus tutorials on the art of soldering and useful programs such as Radio Jupiter Pro and Radio-SkyPipe. These programs will be discussed later in the chapter. Something missing from this image is the RF 2080 calibrator, as this wasn't available at the time the kit was purchased, and was ordered later. The calibrator and its uses will also be discussed later in the chapter, if thinking of purchasing a Radio Jove kit it would be a good idea to order one of these calibrators at the same time, as they are a very useful accessory, and if one is ordered at the same time as the Radio Jove kit there should be only one lot of postage to pay.

Before ordering a Radio Jove kit it is vital to note that the antenna is of a dual di-pole design and will require a reasonably large area to erect it. The two di-poles each measure 7.6 meters (25 feet) in length and are placed at a distance of 6 meters (20 feet) apart.

Depending where in the world the kit is to be used, and the height at which the Sun and Jupiter transits above the local horizon, the height at which the antenna is used can range from 3 meters (10 feet) up to 6 meters (20 feet). If short on space a single di-pole antenna can be used and still be able to receive Jupiter, but a single di-pole antenna will not have the same gain. A single di-pole antenna can however easily pick up radio emissions from the Sun. The Radio Jove website is a good place to start for specific questions on the subject as there is plenty of useful information there to be explored in addition to a list of mentors who may be emailed directly with questions. An electronic newsletter is also produced two or three times a year and can be downloaded from the Radio Jove web site. The newsletters are full of useful information about what is happening in the world of Radio Jove, and contain images sent in from Radio Jove users all around the world. There is also an excellent online server sign up to receive email messages from other Radio Jove users around the world. This server can be of great interest, and shows how other Radio Jove users in different countries have received the same radio emission, plus any updates to software that are available. Occasionally the Radio Jove team run a phone-in where Radio Jove users can call in and discuss all aspects of the Radio Jove project with the Radio Jove mentors.

7.2 A Guide to Building the Radio Jove Receiver

Building of the Radio Jove receiver is estimated to take approximately 9 hours, but this is only an estimate and it may take longer or shorter depending on the builders capabilities. Looking at the manual there is a list of resistors, capacitors, inductors, integrated circuits, transistors, and the other items such as power connections, etc. Images of all the components are included. Within the manual are two columns, one is for parts identification and the other is for part installation, and each one should be ticked after identification and installation has taken place. This may sound like stating the obvious, but if this is done religiously, then if having to stop for whatever reason, for example the telephone rings, it can be easy to pick up and continue from where left off.

Before starting to identify the components it's a good idea to use small stickers, something along the lines of the size used for price tickets in shops. These can be obtained from any office supply store. Work on only one type of component at a time. The manual lists capacitors first, and although they number up to 44 there are in fact 43. This is because capacitor number 7 is no longer used but is still listed. After the first capacitor has been positively identified fold a small sticker around one of the wire connections and mark it C1, and tick the column within the manual. Carry on until all of the capacitors have been identified and labeled. Then do the same with the resistors, of which there are 32. Start with R1 and carry on until all of the others have been identified and tagged. Please note that resistor number 32 does not fit on to the circuit board, and be sure to keep this component separate and in a safe place for the moment. Do the same labeling of the other components, until each and every one has been positively identified and had the corresponding box in the manual ticked. It will be noticed that there is a large silver colored, rectangular shaped object within the components. This is a 20 megahertz crystal, and this should be soldered in place with the other components on the circuit board. The purpose of this crystal is to tune the receiver, and it can only be fitted in one position so there will be no problems with it orientation.

This may come across as rather time consuming, but when soldering each part into location it's just a matter of looking for the right number on the components sticker, and this also acts as a double check if forgetting to tick the column within the manual.

There were two components which identification was a problem. These were zener diodes ZD1 and ZD2 as their marking on the case were slightly different than the marks shown within the manual. ZD1 was marked within the manual as 1 N753 but the diode case read F53A, and ZD2 was marked in the manual as 1 N5231 but the diode case read F231B. This is not uncommon when building projects from kits, as different manufacturers have different markings or even different colors for the same component, but a quick search on the internet soon sorted out which one was which.

Notice also within the construction manual the term "jumper wires" will be used, these are just short lengths of connecting wire that are soldered on to the circuit board to link different parts of the circuit together to complete the electrical circuit in a convenient place.

Looking at the aluminum enclosure, don't assemble this yet, but notice the edges are quite sharp from where they have been cut on a sheet metal cutter. Within the kit there should be a small piece of sandpaper, and it's a good idea to give each edge a light sanding to remove these sharp edges, this will save fingers from being cut on assembly of the enclosure. After sanding it is best to wipe each piece with a clean damp cloth to remove any aluminum dust or shavings, as these can cause short circuits if any get in contact with electrical components. These can then be dried and put away for use later, but leave out the front and rear panels. The front panel has two large holes and one small hole, and the rear has four large holes and one smaller hole. Included within the kit are front and rear decals, these can be peeled off the card and fitted to each panel, taking care to avoid trapping any air, as this can leave unsightly bubbles. After fitting the decals these panels can now be put away with the other parts of the enclosure for use later.

It is time to start soldering in the components, but first get the circuit board in the right orientation. Handle the circuit board on its edges like a DVD, as oils within the fingers can stop solder from forming a good joint. Look for the Radio Jove 20.1

7.2 A Guide to Building the Radio Jove Receiver

Fig. 7.2 Rear view of finished circuit board. (The crystal can be seen at the bottom center of the image)

megahertz receiver printed on the circuit board, as this indicates the top of the circuit board and the front edge (see images below). Look at the holes where the components will be fitted, the circuit board is marked with the number of the component that fits in each hole, and the polarity of polarity-sensitive components are also indicated.

Components like resistors and ceramic capacitors will need their connecting wires bending so they can be fitted into the circuit board. Don't be tempted just to bend them, as if they are bent too close to their ceramic body this can damage the ceramic and this could cause the part to fail, either immediately or at a later stage. It is better to use a pair of needle nose pliers to hold the connecting wire near the ceramic body to support it, and then bend the connecting wire with the fingers. It's also a good idea when bending the connecting wires of an electrolytic capacitor so it can be fitted on to the circuit board not to let the connecting wires from the capacitor come in contact with the case itself. Although they should be insulated it is not worth the risk.

Start by soldering in the resistors first, as they can best put up with the heat of soldering. After each component has been soldered into place turn the board over and trim the excess wire off with a sharp pair of cutters. It is a good idea to have a little container handy to drop these into, as there will be around 200 of these left after building the receiver and they have the habit of getting everyway like needles from a Christmas tree (and are just as unpleasant if not worse to stand on). The last thing to be fitted to the circuit board is the integrated circuits. These can sometimes require a slight bending of the ICs contacts so they can be fitted into their IC socket, but be very careful fitting these as they are easily damaged. After finishing soldering in all the components, it should look like the image (Fig. 7.2).

Fig. 7.3 The circuit board fitted into its enclosure after the final connections have been made and the temporary R32 (51 ohm) has been fitted

It is now time to fit the antenna input socket, power input socket and audio output sockets to the rear panel. This is just a case of putting them through the pre-drilled holes and applying a locking washer and a nut. These only need slight pressure applying to the nut, with a pair of pliers to hold them secure. The circuit board is now fitted to the base of the enclosure. It is mounted on spacers to stop the bottom of the circuit board coming in contact with the aluminum base and causing a short circuit. The rest of the enclosure needs to be built now. This is just a simple case of fitting four channels to one of the ends, and fixing them in with screws, and then sliding in the bottom, front and back panels and securing them with four screws on the other end. The top is left off at this stage as the connection to the antenna, power supply and audio inputs and outputs need to be made. The two control knobs for power/volume and tuning can now be fitted. The resistor number R32 (51 ohm) is now temporally soldered onto the antenna input connection. The purpose of this resistor is to simulate an antenna being used with the receiver, by producing a "dummy load" on the receiver circuit, and this will be removed after testing and tuning of the receiver itself. Once this has been done it should look like the image (Fig. 7.3).

The receiver is now ready for testing and tuning, and this can be done in several different ways depending on the level of equipment available for use. First, a suitable power supply will be needed to power the receiver.

Power Supplies for the Radio Jove Receiver

If unhappy or unsure about using mains/grid voltage, ask the advice of a qualified electrician first. A 12 volt power supply is needed for the Radio Jove receiver, and this can be a battery, such as a car battery. Either use the battery on its own or working through the vehicle's cigarette lighter socket, NOT with the engine running.

If mains/grid power is to be used beware that a well-regulated transformer is needed, not the cheap ones used for running games consoles. Batteries produce power through a chemical reaction and therefore produce true direct current (DC).

A transformer works off an alternating current (AC) and steps it down from mains/grid voltage to a smaller voltage, in our case 12 volts to run the Radio Jove receiver. Many of the cheaper transformers' outputs have very poor rectification properties. If a display of the output of a battery or a well-regulated transformer were to be displayed on an oscilloscope a straight line would be seen, on the other hand if the display of the output from one of the cheaper transformers were seen on the oscilloscope rippling, or in some cases very big spikes, would be seen on the display.

This type is alright for running games, etc., but NOT for the Radio Jove receiver or other radio receivers discussed here.

As these spikes and ripples within the power supply introduce noise within the receiver's electronics. This is a bad thing, as the signal we want to receive is a form of noise, it makes sense not to artificially introduce noise into the receiver. Steer well clear of those transformers that have a voltage selection switch on them to enable different voltages to be selected for different items. It has been found through experience that the voltage marked on the transformer has nothing whatsoever to do with the voltage coming out! After blowing an LED light cluster by setting the transformer to the so-called correct voltage and polarity marked on the side, a quick check is always made of the output of these variable transformers with a voltage check on a multimeter first. This ensures that the voltage and polarity are correct. This simple precaution can save money in the long run.

Testing and Tuning the Receiver

Before switching on the power make sure the polarity of the power supply is correct. The inner part of the power connection is positive and the outer part is negative. This can be done by doing a quick check with a multimeter set on the voltage setting. Apply the test probes to the output of the power supply, with the positive probe at the center and the negative on the outer part, the voltage reading should be 12 volt. If the reading is -12 volt then the polarity is wrong and must be changed before it is applied to the receiver, or the receiver will not work or be permanently damaged. Once the power supply has been correctly identified, it can now be plugged into the power input socket of the receiver.

Fig. 7.4 Testing of the Radio Jove receiver. Using the oscilloscope method and externally powered speakers

The receiver must be "tuned" before it can be used. There are several ways of doing this depending on the level of test equipment that is available. If no test equipment is available to tune the circuit, it can be done by listening to the tone of the receiver and adjusting the tuning until a constant pitch is heard, or by using the program Radio-SkyPipe to chart the changes in the output of the receiver. It will be possible to see any drifting in the output of the receiver on the graph produced by the Radio-SkyPipe program on the computers screen, and changes to the receiver can be made accordingly. Another method is to use a multimeter that is capable of measuring audio frequency voltages to measure the output from the receiver itself. The last method of tuning is to use an oscilloscope to monitor the output. This gives a visual reference to watch to see how the output changes. Please see Fig. 7.4.

It is also a good idea to have either headphones or externally powered speaker(s) to hand. After choosing the preferred method of tuning, and after all necessary test equipment and headphones or speaker(s) have been fitted, it is time to switch on the receiver for the first time. Don't put the headphones on just yet. There should be a high-pitched noise coming from the headphones or external speaker(s). This is the noise being generated within the circuit from the 20 megahertz crystal mentioned earlier.

It would be a good idea to run the finished receiver for about 5 minutes before starting the tuning process, as this will allow the receiver circuitry time to "warm up" and stabilize, this will lead to a more accurate tuning.

It is now a matter of adjusting four components to tune the receiver. First, set the tuning knob on the receiver to the 10 o'clock position, and using the plastic tools included within the kit adjust the component L5 (variable inductor) until a loud

7.2 A Guide to Building the Radio Jove Receiver

Fig. 7.5 The finished Radio Jove receiver front view

tone can be heard. Once this is done forget L5 and move on to components C2 (variable capacitor), C6 (variable capacitor) and L4 (variable inductor). These need to be adjusted until the maximum signal strength is produced. When adjusting the components only make very small adjustments at a time, being careful not to force the adjustment screw in either direction as this can damage the internal parts of the component itself. This may sound a little fiddly, but in fact it is quite simple, especially if using the oscilloscope method, as the actual change in strength of the signal can be seen visually. Not everyone will have access to an oscilloscope but there is no reason why the other tuning methods shouldn't work just as well.

Once the tuning procedure has been completed, and with the power to the receiver switched off, the resistor R32 (51 ohm) can be unsoldered. This resistor must be kept safe in case this tuning procedure needs to be carried out again. A good idea is to stick this resistor to the inside of one of the panels of the enclosure with electrical tape in order to know exactly where to find it if it is needed again. There is one last thing to do before fitting the top to the enclosure, that is to cut through the number 6 "jumper connection" wire. Don't remove it, just snip through the wire and bend it up so it cannot come in contact with any other part. This removes the tuning crystal from the circuit now that it has completed its job of tuning the receiver. By not removing the "jumper" wire, if the tuning process needs to be repeated for whatever reason the "jumper" wire can soon be reconnected. With all of the above tasks completed, the receiver is ready for use, as shown in the Figs. 7.5 and 7.6. Please note the connections on the rear view of the receiver, in particular the two audio outputs, this is so one output can be used with headphones and the other can be used to connect the receiver to a computer or other device.

Fig. 7.6 The finished Radio Jove receiver rear view. Please note the connections, (from left to right) Antenna input, audio 1, audio 2 and 12 volt power input

7.3 Building the Radio Jove Antennas

The antenna for the Radio Jove receiver is a dual di-pole design. This dual di-pole design gives greater gain then a single di-pole, and with the use of a phasing cable and the height at which the antenna is positioned the center of the beam can be controlled to give the best gain possible for the height of the object, be it the Sun or the planet Jupiter, as it appears above the local horizon.

The kit includes six ceramic insulators, a coil of copper wire, toroid collars, a coil of RG59/U antenna cable with "twist on" F connector fittings and a power combiner. The power combiner is a connection with two inputs and one output. This allows the two di-pole antennas to be coupled into the two inputs and the single output then goes to the Radio Jove receiver. A di-pole antenna is an antenna cut to half the wavelength of the in-coming radio wave that is to be received. In the case of the Radio Jove receiver working at 20.1 megahertz this half wavelength distance is 7.09 meters (23.28 feet). Also within the kit are six ceramic insulators, and the job of the insulators is to allow the antenna wire to hang in free space from the supports, without allowing the antenna wire to come in contact with either of the supports or the antenna cable. This is done by fitting an insulator at the end of each of the antenna wires, plus one in the center of the wire where the antenna cable is connected to the antenna wire itself.

The first task is to unroll the coil of copper wire which is included within the kit and cut it into the correct lengths to make the two di-pole antennas. It can be quite

7.3 Building the Radio Jove Antennas

awkward trying to measure the correct lengths of wire while it is in one long coil as it will have the tendency to keep rolling back on itself like a roll of wallpaper. A good way to cut the antenna wire to save on measuring out long lengths of wire is to cut it in half and then half again to have four equal lengths of wire. These will still be too long to make the antenna, but the length allows enough wire to pass through the ceramic insulators and twist it back on itself in order to secure the wire to the insulators.

When the four equal lengths of wire have been cut, start to fix them to the insulators at the correct length. A good way to get this length is to measure out the correct length on a floor and mark this length with two chalk marks, this will save trying to hold the antenna wire and stop it coiling back on itself and at the same time trying to hold a tape measure. It would help to have assistance with this part. Holding one insulator on a chalk mark and an assistant holding one insulator on the other chalk mark, then getting hold of one of the lengths of antenna wire, each person can allow the same amount of wire spare at each end to allow this to be passed through the insulator. After passing the wire through the insulator, twist it back on itself. Don't go right up to the insulator, as if it is fitted too tightly onto the ceramic insulator it may cause it to crack or the wire may become damaged rubbing against the hard insulator when the antenna moves in the wind. Once this task has been performed there should be two antenna wires of the correct length with an insulator at either end and one in the middle. Don't solder these in place just yet.

The two antennas can be carefully coiled up and put somewhere safe for the time being. When coiling the antenna wires, try not to let the ceramic insulators knock together as they can soon have their protective glazing damaged. Before discussing the cutting to length of the antenna cable, here are a few points to follow when handling the antenna cable. The antenna cable has a natural twist already formed within the cable itself. When coiling these types of cable allow it to follow this natural twist, and don't make the loops too small. This will allow the cable to be coiled and uncoiled more easily as opposed to fighting against its natural twist, and the central conducting wire, which is solid in this cable, will not become stressed and so will be less likely to break. It looks neater, and the antenna cable doesn't take on the look of an old bird's nest. Under no circumstances allow the antenna wires or the antenna cables to become kinked. If needing to take the antenna cable round a corner, do so with a gentle curving of the wire rather than a sharp bend. This may sound like belaboring the point, but if this practice and these few simple precautions are followed, there is no reason why the antennas shouldn't provide many years of service.

A good sharp pair of wire cutters will be needed to cut through the antenna cable. Nothing comes across as more amateurish then a length of cable which looks like it has been hacked through with a blunt pair of scissors. This can cause all sorts of problems when trying to fit the connections later on. The antenna cable needs to be cut at set lengths so both of the di-pole antennas will work correctly. The lengths of each cable are cut in factors of the wavelength of the signal that the Radio Jove receiver is going to receive. There are four cables to cut, so it is a good idea to wrap a different color of electrical tape around each of the cut lengths and make a quick note of which color corresponds to which length. This will save time later, as the cable will only need measuring once just before cutting, as it is hard to judge the lengths of cables when they are coiled up.

First, cut the two cables which come from the di-pole antennas. These are of equal length equivalent to one wavelength using the properties of this particular antenna cable supplied. If other cables are used these lengths may be subject to change. Don't mix different types of antenna cables, as this will cause an impedance miss-match. This first measurement is 9.58 meter (32.31 feet). All the cables must be cut to their exact measurement. Now cut the phasing cable. This is cut to 0.375 of the length of the wavelength, and this equates to 3.69 meters (12.12 feet). The last cable to be cut is the cable that comes from the power combiner to the receiver and is 0.5 of a wavelength, which is equal to 4.93 meters (16.16 feet). After cutting all four cables it may be found there is an inch or so of cable left over. Do not throw this away as it can be used to connect the calibrator to the receiver. A length of 50 millimeters (2 inches) is all that is needed for this.

Next get one of the lengths of cable that will be fitted to the di-pole antenna wire. Slide three of the toroid cores supplied within the kit on to the end of the antenna cable which will be fitted to the antenna wire. These improve the antenna performance and help stop interference entering the open end of the cable. These can be a tight fit, but they should be easy enough to push on. Try not to knock these together or force them on as they can be easily damaged. Slide them down around 500 millimeters (2 feet) and apply a little electrical tape to temporally hold them in place while soldering the connections. Now, strip the outer sleeve of insulation off the end of the antenna cable, to a length of 100 millimeters (4 inches). Take care not to cut into the braided layer underneath. Two or three pieces of braiding will probably be lost, this is normal as long as it doesn't run into half a dozen or more.

Now carefully untangle this braiding off the inner sleeves insulation. This can be done using the tip of a small screwdriver. Again the tip of a ball point pen can be used for this as this is a nice round end with no sharp edges to it. Once this has been done, twist together all the strands to make one wire. This will be one of the connections. Next, cut off 50 millimeters (2 inches) of the inner insulation sleeve to expose the solid central connection. Be careful not to score the central wire with the knife as this will cause the wire to have a weak spot which could be damaged when the antenna is blown about by the wind. It is now time to start making the connections to the antenna wire. First place the antenna cable over the central insulator and the antenna wire, and hold it in place with a couple of cable ties. Please see Fig. 7.7.

Once the antenna cable has been secured, the connections need to be made from the antenna cable to the antenna wire itself. Start with the braided wire. Twist this around one of the antenna wires on one side of the central insulator, but don't pull the wire too tight, a little slack should be left to allow the antenna to move in the wind. Next, fit the solid central wire to the antenna wire on the other side of the central insulator. Special care must be taken here, with a gentle curve of the antenna cables central connection bring it up to the antenna wire, and allow enough slack in the antenna wire to allow it to move without undue pressure being applied to the solid central antenna wire, for the same reason as above.

7.3 Building the Radio Jove Antennas

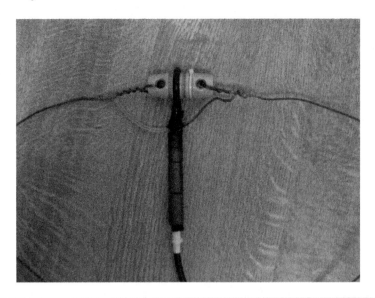

Fig. 7.7 The connection of antenna cable on the central insulator of one of the di-pole antenna. Note the three toroid collars seen on the antenna cable below the central insulator

Solder the wires in place. A more powerful soldering iron will be needed, one that is somewhere in the region of 75 watts will do the trick, because of the thickness of the wire used for the antenna. Don't try and use the 25–30 watts soldering iron used to build the Radio Jove receiver circuit board, as this will not produce adequate heat input into the antenna wire, and any melted solder will just sit on the antenna wire and produce what is known as a "dry" joint. Enough heat must be applied to allow the solder to flow freely through the antenna wires, including the wires from the antenna cable. Once the antenna cable has been soldered onto the antenna wire, solder the twisted antenna wire on each of the insulators at each end of the antenna wire to stop them from coming free. After the soldering has been completed, slide the toroid collars back up to the top of the cable and hold them in place with a cable tie, as shown in the above image.

The open end of the antenna cable must now be waterproofed to keep out moisture and dirt. One way of doing this is to apply electrical tape to the open end of the antenna cable and then apply a waterproof coating to this. This can be done using waterproofing specially made for coaxial cable, but instead an automotive ignition sealer can be used; not the type used to dispel moisture, but the type that leaves behind a thin waterproof skin. Spray a liberal amount all around the open end of the antenna cable, it doesn't matter if the spray covers other parts such as the insulator. Allow this first coat 10–15 minutes to dry, then apply a second coat of the spray. This should be adequate to waterproof the cable. Depending on annual rainfall if this task is repeated two or three times a year when the joint is dry, this should be enough to top-up the waterproofing action of the spray. If this is done a can of ignition sealer will last for years. A word of caution: don't use the spray indoors as it has a terrible chemical odor to it, which will fill every room in the home.

The other di-pole antenna can then be constructed in the same way. Once both di-poles have been finished the "F" connectors will need to be fitted to the ends of all the antenna cables. Start by removing 25 millimeters (1 inch) of the outer sleeve of insulation from the antenna cable, remembering to be careful not to damage the braiding underneath. This time only untangle half the length of the braiding and fold this back on itself over the un-braided part. Remove the insulation from the solid central conductor, being careful not to score the central conductor itself. It is now time to fit the "F" connector. Put on the connector with a clockwise twisting motion, apply firm pressure until the threads on the inside of the connector "bite" and the connector will screw itself home. Hand pressure should be enough to fit the connector, and there should be no need to force it on with pliers, although it can get a little tight towards the end and a small amount of pressure can be applied. Once all the "F" connections have been fitted to the antenna cables, wrap a little electrical tape around the end that fits on to the cable itself. Two or three turns should be enough, starting with the tape half on the "F" connector and half on the outer sleeve of the insulation of the antenna cable. This is just a precaution to stop dirt and damp entering through the gap between the connector and the outer sleeve of insulation and corroding the antenna cable. Different colored tape could be used to indicate which cable is which, as this is useful to know until becoming more familiar where each cable fits.

This completes the construction of the antennas, meaning now it is time to fabricate some antenna supports. The antennas are used horizontally polarized, and need to be mounted at a height which will be suitable to get the object, either the Sun or Jupiter, as close to the center of the antenna pattern as possible. The height at which the antennas are mounted can be anything between 3 meters (10 feet) minimum to 6 meters (20 feet) maximum depending on where the antennas are used in the world and the altitude at which the Sun and Jupiter appear in the sky above the local horizon.

Four supports will be needed from which to suspend the antennas, and guide ropes to help steady the supports will also be needed. The supports can either be made of plastic tubing, wood or metal. Plastic is cheap, corrosion resistant, light weight, and can be carried in sections to an observing site and assembled there, but at long lengths plastic tubing can become very flexible and prone to splitting or breaking.

Choosing to use plastic tubing will probably be limiting the antenna height to the lower levels.

Wood is a great all-rounder. It's relatively cheap to purchase, but in long lengths can prove a little awkward to handle. If a permanent site is in mind then wooden masts permanently fixed into the ground and supported with guide ropes would be great. The only maintenance needed would be the application of a timber preservative once or twice a year. If the antenna supports are permanently sited, it would be a good idea to engineer some sort of pulley system that could remotely alter the height of the antennas from the ground. Metal supports are more expensive than wood or plastic, and they are heavier, and prone to corrosion, but reasonably rigid at longer lengths, especially if supported with good guide ropes.

One idea is to use metal washing line props as these are very useful as antenna supports. These are available from most DIY/hardware stores and are quite cheap to purchase. They are telescopic and can be collapsed down like a whip antenna on a car for easy storage and transport. When folded down they are no longer than about 1 meter (39 inches) in length. They are alright for making observations where the antenna is used up to 4.5 meters (15 feet). Above this they are unsuitable and more rigid supports are needed. Eight of them will be required, cut off the plastic hook at the end of four of them, leave the hook on the other four this is ideal for hanging the antenna. Next, taking the four poles with the plastic hooks still attached, remove the plastic bung from the bottom of each pole, so the thinner of the two poles from the one which has had the hook cut off can be slid inside. Two poles can be easily fitted together to form four longer supports. The poles can be easily drilled to accept fixings for guide ropes. A good idea is to make hoops to fit into the holes in the poles rather than passing the guide rope through the holes in the poles, as any rough edges on the inside of the drilled hole will act as a saw and cut through the guide rope by the motion of the poles moving in the wind. These hoops can be made simply by bending some small diameter metal into a triangular shape and fitting the two open ends inside the poles. These could be made using old round tent pegs or wire coat hangers.

These poles, if erected properly and equipped with guide ropes and checked from time to time, can cope with quite bad weather conditions. Sometimes these poles come with some kind of plastic protective coating, so no painting is required. The quality of this protective coating isn't great, but they should last quite a few years before the corrosion sets in.

A point worth mentioning about guide ropes. Washing lines are fine for this as they are cheap and come in long lengths, but some come with a metal core surrounded by a plastic sleeve or coating. These must be avoided, as the metal within the rope will disrupt the shape of the antenna pattern, as will any metal nearby, such has a chain link fence (the above metal antenna poles will not affect the antenna pattern). Avoid ropes made with natural fibers, these can soon rot if they are constantly getting wet, and they can also break down by exposure to ultraviolet radiation from the Sun unless they have some sort of ultraviolet protective coating. The best rope for this purpose is nylon rope, which is light weight, very strong, and because it is a washing line and therefore designed to be used outside it will have some kind of ultraviolet coating applied to it to protect it from breaking down in sunlight.

When cutting nylon rope, a good idea is to melt the fibers of the cut ends with a gas cooker ring or candle to prevent the rope from fraying. Guide ropes can be anchored to the ground using metal, wooden or plastic tent pegs. Avoid the small round ones as mentioned earlier, these small round pegs will be forever coming lose and needing to be hammered in again. Use the better quality triangular shaped ones; these are more expensive but worth it in the long run.

7.4 Antenna Configurations

To understand how the antenna works we will use the example of the antenna pattern produced by a single di-pole antenna first. Looking at the Fig. 7.8, the antenna is the line in the center of the ellipse; the ellipse shape is the antenna pattern. This shape would be of a three dimensional pattern as if the antenna was passed through the center of a huge jelly doughnut.

Figure 7.8 shows the maximum gain is in the center of the antenna where the central insulator is situated, the minimum gain is at the ends of the wire itself. The ends of minimum gain can be used to an advantage, if for example there is a constant source of interference, move the antenna so the minimum gain is pointing in the direction of the source of interference. This should help with the interference problem. The antenna for the Radio Jove receiver has many different configurations, but all have the antenna horizontally polarized. The configurations are designed to get the maximum gain out of the antenna when Jupiter or the Sun are at different altitudes in the sky as seen from the local horizon.

The configurations described here will be suitable for most mid latitudes in the northern hemisphere. If in the southern hemisphere all that is required is to change round the phasing cable so the antenna pattern will be north phasing. The phasing cable has the effect of producing a larger gain on the south facing antenna pattern with a more elongated shape, also it has the effect of lowering the altitude at which the maximum gain is achieved.

Figures 7.9, 7.10, 7.11, and 7.12 are screen shots from the computer program "Radio Jupiter Pro" (this program will be covered in more detail later). It will be easier to explain the antenna patterns and the effect of the height at which they are used by reference to these screen shots. Looking at the first Fig. 7.9, the antenna pattern is the ellipse shape in the middle. The two horizontal lines in the middle represent the two di-poles, and the cross in the very center is the point of maximum gain of the antenna in this configuration. Jupiter can be seen just inside the antenna

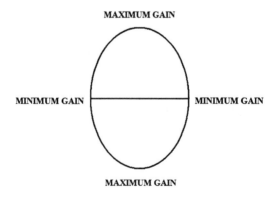

Fig. 7.8 Antenna pattern of a single di-pole antenna

7.4 Antenna Configurations

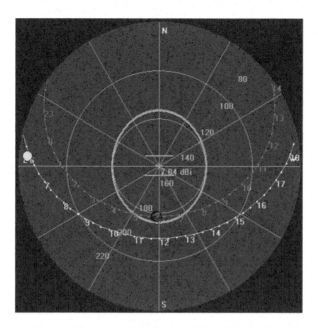

Fig. 7.9 Dual di-pole without the 135 degree phasing cable. At a height 3 meter (10 feet)

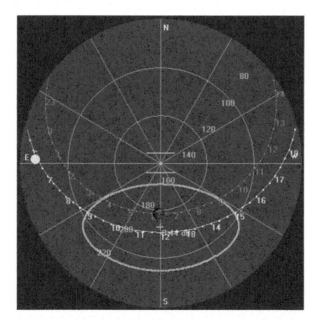

Fig. 7.10 Dual di-pole antenna with 135 degrees phasing cable, south phasing, at the height of 3 meter (10 feet)

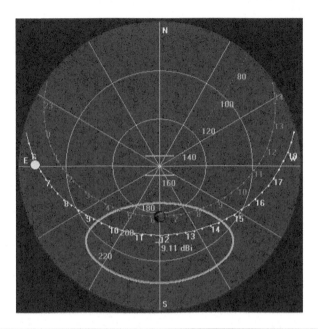

Fig. 7.11 Dual di-pole antenna with 135 degrees phasing cable, south phasing, at the height of 4.5 meter (15 feet)

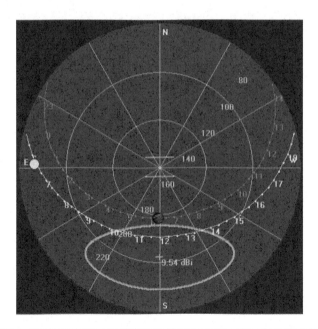

Fig. 7.12 Dual di-pole antenna with 135 degrees phasing cable, south phasing, at the height of 6 meter (20 feet)

7.4 Antenna Configurations

pattern bottom center and the Sun can be seen on the left hand side. The Sun's track across the sky is the line below the antenna pattern, but in mid-summer if the Sun reaches 60 degrees or more above the local horizon during transit this configuration would work well.

Looking at the next Fig. 7.10, this has the 135 degree phasing cable in place. Nothing else has been changed; the antenna di-poles are still the horizontal lines in the middle of the image. But look how the shape of the antenna pattern has been changed by fitting the phasing cable. Notice the cross in the center of the ellipse, where the maximum gain of the antenna is in this configuration, the maximum gain is at 50 degrees above the local horizon, but look how close it is to Jupiter. This antenna configuration would be suitable for receiving radio emissions from the planet. Also notice the track below Jupiter, that is of the Sun. The configuration would also work very well to tune in to the radio emissions from the Sun.

Looking at Fig. 7.11, the only variable that has been changed is the height of the antenna, which is now 4.5 meters (15 feet). In this configuration the maximum gain will be at 40 degrees above the local horizon. Look how far Jupiter has been moved away from the center. This change could make all the difference to either receiving radio emissions or not. The Sun fares slightly better in this configuration at this time of the year (August).

The next Fig. 7.12 is of the antenna pattern with the antenna set at a height of 6 meters (20 feet). At this height the maximum gain from the antenna will be at 30 degrees altitude above the horizon. If Jupiter is low in the sky, as it has been in the last few years, this antenna configuration would get the most gain from the antenna. As the Sun makes its familiar rise and fall in altitude in the sky over the course of a year, from its highest position in midsummer to its lowest position in midwinter, the height at which the antennas are used will need to be adjusted, to keep the object, either the Sun or Jupiter, as close to the center as possible to get the most gain from the antenna.

Why must the object be kept at the center of the antenna pattern to get the most gain? Surely if the object is within the antenna pattern the receiver should be able to pick up the radio emissions. To explain this, switch on a flashlight inside a dark room and there will be directly in front of the flashlight a fully lit area of light projected on to a nearby wall, but all around this fully lit area will be a semi-lit area. Think of this semi-lit and fully lit area as the antenna pattern or beam. If we take Jupiter as the example, when Jupiter is completely outside the antenna beam we can't see it as its in darkness, neither can the receiver, but as the Earth rotates and brings Jupiter across the sky into the antenna beam it first enters the semi-lit area, so in theory the receiver could pick up radio emissions from the planet but the gain from the antenna will be lower, and as Jupiter makes its way into the fully lit area of the beam this is where we get the most gain and this is the best time to listen for radio emissions.

As the rotation of the Earth now moves Jupiter out of the fully lit area on the opposite side of the antenna beam, we start to lose the gain of our antenna once again. Note that the ellipse used to indicate the antenna pattern is a theoretical depiction of the antenna pattern and it is not set in stone. It is therefore best to start listening an hour and a half to 2 hours either side of the transit time marked on the antenna pattern. This is the best chance to pick up the radio emissions. There is a way around this problem by using a different antenna design, and there are a number of different types available on the market, such as the steerable high gain antennas like a three element Yagi. But this will bring its own problems as they will need an antenna mast to mount the antenna and a suitable way to steer the antenna, as these type of antenna are highly directional and will need moving every hour or so, not to mention the cost of such an antenna and the steering device.

7.5 Software

Included within the Radio Jove receiver kit are two computer programs, "Radio-SkyPipe" and "Radio Jupiter Pro". Both are the creation of Jim Sky of Radio Sky Publishing, www.radioskypublishings. Radio-SkyPipe is a very useful program and easy to understand and use. Once loaded onto the computer, it turns the computer into a chart recorder with an on-screen display similar to the old paper chart recorders, but without generating miles of paper tracings. The version supplied has some of the utilities blocked, but it is a useful starting point, and this is enough to get the receiver up and running and will give a visual display of the output of the Radio Jove receiver. Depending on the operating system of the computer, some of the input characteristics on the computers audio input settings will need to be set manually, in a similar way as described within the chapter on the SuperSID monitor, or if a more modern operating program used this may be done automatically by the computer. A window will pop up asking if microphone or inline input is required, check the inline box and if possible mute the microphone input.

Soon it maybe found that the supplied version of the program has outgrown its usefulness and the fully functioning version is required. If wishing to purchase the fully functioning program from the Radio Sky website, which is thoroughly recommended, it can be easily downloaded over the internet. Once payment is received an email will be sent with a code, and once this code is entered into the program it unlocks the rest of the program's functions. After purchasing the program, any updates and improvements made to the program are free to download and install, so it is always possible to have the up-to-date version of the program. The fully functioning program has lots of useful functions, far too many to be discussed here, but it is worth pointing out a few. The program is capable of recording sound files, and playing them back, also it is programmable so it can be triggered automatically to record sound if a level set by the operator is exceeded. It has a calibration wizard function that can be used with the 2080 calibrator, in order to calibrate the program to the Radio Jove receiver. The on-screen instructions are clear and easy to follow (this will be covered later in the chapter). Any files will automatically be saved and can be easily retrieved later. Through the program the computer's sound card can be accessed, to change sampling rates, etc.

Once Radio-SkyPipe is running, a small control panel will be shown in the top left of the screen. By simply clicking one of these buttons a large range of tasks can

7.5 Software

be performed, such as changing the scales of the display or a record button which will immediately start recording a sound file. If seeing or hearing something of interest notations can be added to the graph so it can be easily found later, or even save the screen display straight to a file for retrieving later. To sum up, this is a very versatile program and it is very good at what it does, it is easy to use, although some aspects of the program may take a little practice to master.

The other piece of software is one of those things that comes along and makes one wonder how life went on before without it. Radio Jupiter Pro is one of these things. The version supplied with the Radio Jove kit has some of the functions blocked as with the previous program. This is fine as an introduction, though this too will soon be outgrown, and the fully functioning program will be required. This is also downloadable from the Radio Sky website, using the same procedure as described with the previous program. As with Radio-SkyPipe, once the software has been purchased any updates or improvements made to the program are free to download and install, so the latest version will always be available to work from.

The fully functioning Radio Jupiter Pro software has everything needed for observing Jupiter and the Sun using the Radio Jove receiver. Once the software has been loaded there are a few things that need to be set, latitude and longitude, plus preferred date format and time zone the program then does the rest. On opening the program a window pops up showing the relative position of Jupiter and the Sun plus the time and date. At the bottom of the window is a smaller window that informs the user if there is a possibility of any radio noise storms from Jupiter happening at that moment. Click on the next window and the "radio storm prediction" window pops up. This window displays a chart showing the predictions for radio storms for the next 24 hours. There are buttons to skip to the next day or so into the future or the past. The predictions are also displayed in written form, giving approximate start and finishing times. The Sun's visibility above the horizon is also shown. Moving the computer's curser across the screen the details of the time will be seen to change according to the curser's movement, and the positions of the Sun and Jupiter will be shown in degrees of altitude.

The next window is the "Io phase plan". This charts Jupiter and the central meridian longitudes, and gives a probability of receiving a radio storm if Jupiter is in the right part of the sky for the observer. Looking at the Fig. 7.13, a screen shot of this particular window. This image may seem strange at first sight, but it is quite straightforward to understand once a number of key features are pointed out.

Looking at the image, Jupiter can be seen on the right hand side just below center. The lines running from left to right at an angle are Earth times in 24 hours. As Jupiter moves along these lines the probability of picking up a radio storm can be seen to change. Looking on the left hand side of the image there is a color-coded strip, starting with black at the bottom and red at the top, there is also a percentage value marked at different points along the strip to give the user an idea of the chances of picking up a radio storm. Moving the computer curser across the screen the probability can be seen to change on the right hand side of the image.

The Io phase window can be useful, because as Jupiter moves along the time lines, the probability of receiving radio emissions from Jupiter can be seen at any point.

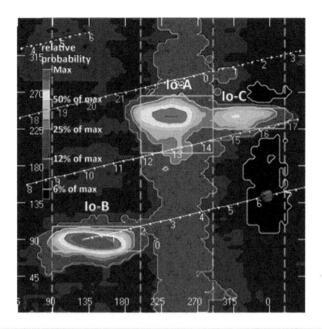

Fig. 7.13 Radio Jupiter Pro, the Io phase plane window

As the planet will be seen to cross the different colored areas, the brighter the color the greater the chances of receiving radio emissions. Ideally Jupiter wants to be in the red area just as Jupiter is passing through the center of the antenna beam. The next window is the antenna pattern window. Recalling the screen shot images from the section covering antenna configurations, this is what this particular window looks like. This part of the program is great for knowing the correct time to listen to the Sun and Jupiter, and gives a visual indication of the positions of the Sun and Jupiter within the antenna beam. The program comes with a number of pre-loaded antennas ready for the user to load the particular one they are thinking of using. This can be good, as it can be seen which configuration is the best for the location, or if coming up with an antenna configuration of their own this can also be loaded into the program. By moving the computer curser around on the map the altitude and azimuth coordinates can be seen to change in the top left hand side of the screen, and this is extremely useful for doing a quick check on the altitude of the Sun and Jupiter.

Choose to personalize the screen, by having stars on the background down to magnitude 5, or another magnitude set by the user to match their own location. This could be useful if having trouble finding Jupiter in the sky. Choose to have the galactic plane shown on the map. This particular part of the sky is rather noisy at radio wavelengths, and contains some of the "brighter" radio sources, as it follows the Milky Way across the sky. This is where Karl Jansky discovered the radio noise coming from the galactic center in the early 1930s from the powerful radio source "Sagittarius A" in the constellation of Sagittarius. The tracks of the Sun and Jupiter are also shown and either one can be switched on or off if one chooses. One good feature about this window is that it keeps updating itself, and the progress of the Sun and Jupiter can be watched through the antenna beam while carrying out radio observations.

The next window, the "altitude vs. azimuth" window, simply shows the altitude and azimuth track of the Sun and Jupiter across the sky. Use this window to carry out checks of the altitude of the Sun and Jupiter, in order to gain the best antenna configuration to get the maximum gain from the antenna. The next window is the "ephemeris" window, and this shows a table giving the predicted positions of the Sun and Jupiter at given intervals and can be set by the user if need be. The next window is the "plot Jovicentric declination of the Earth" window. This window shows the relative position of the planet Jupiter in the sky over any year the user chooses to set. This has been useful in recent years while waiting for Jupiter to return back into the northern hemisphere tracking Jupiter's progress making its way from the southern skies then crossing the celestial equator and making a welcome return into the northern skies again, ready for visual and radio observations.

The next window is the "yearly visibility schedule". This window will calculate the best times to observe the planet Jupiter over a given year. For instance, the schedule will indicate when Jupiter will be above the horizon when the Sun has set. Its graph is quite easy to understand. Jupiter is shown on the graph as a cyan (blue green) color, and the Sun as the color yellow. If both are in the sky together the color changes to a light green, and if neither are above the horizon the color white will be seen. The months of the year run vertically down the left hand side and there are three hourly time lines running from top to bottom. Using this window it can be seen instantly the best months of a year to observe Jupiter. A word of caution, which may be less of a problem nowadays with the speed of the modern day processors on computers, but is still worth mentioning. There are two settings on this particular window, "high resolution" and "low resolution", and if using a reasonably fast computer leave the setting on "high resolution", if using a slower machine use the "low resolution" setting. Trying to use the "high resolution" setting on a slower computer it may take several minutes to produce the graph.

The next window is the "observer's log". This is an especially likable part of the program when listening in person as well as when recording the output from the Radio Jove receiver on Radio-SkyPipe. On hearing a radio emission there is no need to scramble around for a pencil and paper, while at the same time trying to look at the clock or watch to make a note of the time, then trying to write it all down for use later – all while holding a flashlight in the mouth. The "observer's log" does all this, so we can get on with the job in hand and listen. The observer's log window in its simplest form shows three different types of radio storm signals: "S" bursts, "L" bursts and undetermined. Each one is divided into three different strengths: weak, medium, and strong. On hearing a radio emission, for example a medium "S" burst, just click on the medium "S" burst button and the program will do the rest. It will log the type of emission, the time and date, and save information about the position of Jupiter in the sky at the time of the storm. This can then be saved to a file and studied later or printed out to give a permanent copy of the observation.

The final window in this very useful program is the "automated action parameters" window. This allows the user to set specific conditions and tasks for the program to perform, for example to look for a particular type of Jovian storm at a specific date and time, and then make the user aware of this.

7.6 Calibration

Although the Radio Jove receiver will work without being calibrated, and any radio emissions received are still good, to some extent they will be useless as they cannot be compared with any other observations made by other Radio Jove users. To give an example of this, think of the mess if cars were sold without speedometers, it would be impossible to enforce a speed limit. This is why car speedometers are calibrated. This is also true with the Radio Jove receiver. It helps if everyone is using the same unit, as mismatching of units can prove very expensive not to mention very embarrassing for all concerned. A cautionary example springs to mind. Contractors building parts of a Mars Climate Orbiter for NASA made the mistake of employing one company working in imperial units, to design the descent engine to switch off at X number of feet, the other working in metric units programmed the descent engine computer to switch off at X number of meters. So the space probe was three times the height it should have been when the descent engine switched off leaving the space probe to crash on to the surface of Mars. This makes a good case for everyone's Radio Jove receiver to be calibrated in the same way and using the same units.

In radio astronomy the term "antenna temperature" will keep coming up. Antenna temperature is used as a measure of the power per unit of bandwidth of the antenna, and is measured in degrees Kelvin. This can require some serious mathematics to explain fully, therefore a thorough explanation goes beyond the realms of this book. To keep it simple, the antenna temperature is not a physical measure of the antenna temperature itself, but more of a measure of radio energy the antenna is receiving from a radio source. For example, the calibrator discussed below simulates an antenna temperature of 25,000 degrees Kelvin into the Radio Jove receiver. Think of temperature as not being either hot or cold but as a measure of the energy at which atoms themselves vibrate or move. At absolute zero, or zero degrees Kelvin, all the vibrations and movements within the atoms of a material stop. These atoms have no thermal energy left. The Sun's outer atmosphere or corona has a temperature of around two million degrees Kelvin. If it were possible to measure the physical temperature of the corona with a thermometer we would probably freeze to death trying to do so if the Sun wasn't nearby. The Sun's corona is so rarefied and the atoms so far apart there is no physical heat. However, the energy within the movement of these atoms of the corona would be around the two million degree mark.

To calibrate the Radio Jove receiver we need to input a signal of a known quality and power in order to calibrate the Radio Jove receiver to the program Radio-Skypipe. A way to do this is to use the RF-2080 CF. Please see Fig. 7.14.

7.6 Calibration

Fig. 7.14 The RF-2080 C/F, calibrator/filter

This item can be ordered at the same time as the Radio Jove kit. It is supplied ready-built and comes in two different types. The first is the RF-2080 C. This particular model only has the calibration unit fitted. This is a good chose if living in a radio quiet site, but for most of us the RF-2080 CF will prove a better purchase. This particular model also has a narrow band radio filter fitted alongside the calibrator section of the unit. This filter has proven itself useful in blocking some of the radio interference from some observing sites. There is a down side to this filter, a small loss of signal strength from the antenna to the receiver due to the internal circuitry of the filter. The up side is that the benefit of the filter outweighs this small signal loss.

The calibrator is fitted between the antenna and the Radio Jove receiver. When the calibrator is switched on it introduces a white noise signal of a known quality and power (25,000 degrees Kelvin) into the Radio Jove receiver. Using the program Radio-SkyPipe, click onto the tools tab at the top of the screen and scroll down the list, the last on the list is the calibration wizard. Selecting this starts the calibration process. There are a number of on screen instructions all of which are quite easy to follow. Once carried out, the program Radio-SkyPipe will automatically carry out the calibration. When done, unless the volume control on the Radio Jove receiver is altered, the calibration will hold fast. If the volume control is moved, then the calibration wizard will need to be run again. This is why a volume control on the headphones is desirable, as not to disturb the calibration. Once calibrated to the Radio-SkyPipe, it will be possible to gain an insight to whether the observing site is "radio" quiet or not.

As mentioned above, the calibrator simulates an antenna temperature of 25,000 degrees Kelvin. This is generally considered to be a radio quiet site. Operate the switch on the calibrator this will switch off the calibrator and switch back to the antenna itself, if there is not a large increase in noise then this is a radio quiet site in which to observe, and there should be no problem in receiving radio emissions from Jupiter or the Sun. If there is a large increase in noise when switching back to the antenna, this indicates a noisy radio site. Don't worry, all is not lost. If a trace on Radio-SkyPipe is started this will give an indication of the level of background noise at the observing site.

A good idea is to leave the Radio Jove receiver and Radio-SkyPipe running over the course of a few days and a pattern within the trace itself may start to be seen. Look for times when there are few spikes on the graph, these are interference or from lightning. Interference can come from almost any electrical device, but some are worse than others. It has been found that the interference from such electrical devices reduce from around midnight to about 5 am, when everyone has switched off their television sets and gone to bed. It is not uncommon to get the odd spike of interference now and again from heating thermostats, especially on cold nights. Having a background level which is constantly above 750,000 degrees Kelvin means that there will be very little chance of picking up the planet Jupiter. The radio storms from the giant planet don't exceed this level, try and find a quieter site for the antenna. Other Radio Jove users have found moving their antenna only a short distance from its original location and this has been enough to make all the difference. Either way, it should still be able to pick up the radio emissions from the Sun as these can reach levels of several million degrees, depending on the level of background interference, some of the weaker radio emissions from the Sun maybe missed.

If lucky enough to have a quiet site, run the Radio Jove receiver over a number of days, it may be noticed that there is a rise in the background levels on the Radio-SkyPipe graphs. This will appear to be happening roughly at the same time each day. On closer inspection it may be noticed that this is happening 4 minutes earlier each day. This rise in the background level is the same phenomenon Karl Jansky picked up with his merry-go-round antenna in the early 1930s, the galactic center in the constellation of Sagittarius. One Radio Jove user has even picked up a pulsar using their equipment.

7.7 Radio Emissions from the Sun

The first person recorded to try and detect radio emissions from the Sun was Thomas Alva Edison. This is the same Thomas Edison famed for many inventions, such as the light bulb, phonograph, the telephone transmitter and the electric chair. Edison tried to detect radio emissions from the Sun in the early 1890s with his laboratory assistant. Descriptions of the construction of the antenna record that it was made up of coils of wire wrapped around a metallic core, possibly one of wrought or cast iron, but no records are known to exist of any radio emissions having been received. The reasons for Edison's failure to receive radio emissions could be due to a number of factors, such as the possibility that he was trying at the wrong frequency or that his

7.7 Radio Emissions from the Sun   161

equipment wasn't sensitive enough for the job. It could be that he chose to try it at solar minimum, when the chances of receiving radio emissions were at their lowest. Whatever the reason for the failure, Edison didn't take it any further.

The next recorded person to try was Sir Oliver J. Lodge. He made his attempt between 1895 and 1900. He had built himself a more sensitive receiver and antenna then Edison had used, but even this wasn't sensitive enough to detect the Sun's radio emissions. There are records of other observers who tried and failed over the years to receive radio emissions from the Sun. It wasn't until the existence of the ionosphere was proven in the 1920s, when its effects on different radio frequencies was starting to be understood, that it dawned on observers that they needed to use the correct frequency, i.e. one that the ionosphere would allow to pass through it in order to pick up radio emissions from the Sun.

By now the 11-year solar cycle was beginning to be understood, but as yet no one had made the connection between solar activity and radio emissions. Skip forward to the 1940s and the outbreak of World War Two and the first attempt by the British to use their new invention of radar. In 1942 the British had built up what was known as the "chain home stations-radar cover". This consisted of a large number of radar instillations that covered the entire length of the east coast of the United Kingdom, from the most northern part of Scotland all the way down the east coast and most of southern England. Some of these radar stations reported receiving a strong signal that seemed to be jamming the operation of the radar at particular instillations. The signal seemed to be coming from the east, over mainland Europe. This was a cause of great concern, as it was first thought that the enemy had managed to develop a new piece of equipment that was powerful enough to produce a signal capable of jamming some of the British radar instillations. This would have made the radar all but useless at warning against an attack from the air.

J.S. Hey, whose wartime job was to be part of the operational research group studying the effectiveness and efficiency of the new radar system, undertook a detailed examination of all the radar equipment. When they found it was working correctly their attention then turned to the signal that had been received. They found that at the time the jamming signal had been picked up the Sun was positioned low in the sky over mainland Europe. This lead Hey to come up with a theory that the jamming signal wasn't from the enemy but, could be radio emissions coming from the Sun. Astronomers observing the Sun had noted that a large group of new sunspots had recently appeared on the Sun's disc. Hey's theory was correct. What the radar operators had in fact received was radio emissions from the Sun caused by the increased solar activity. This then made the link between solar activity and radio emissions from the Sun.

Unlike the planet Jupiter, where radio emissions can be predicted with a certain amount of accuracy, the Sun is a law unto itself. Take for instance the lead up to the 2012 solar maximum, which was very slow in coming. After observing the Sun for weeks during the summer of 2011 observers saw very little in the way of activity. It has now been suggested that the 2012 solar maximum will not happen until late 2013 or early 2014, and solar astronomers have gone as far as saying the next solar maximum, due around 2024, may be even worse; time will tell. The Sun has done strange things in the past. For example, the Maunder minimum around the year 1715, where solar activity was low for about 70 years. It has been suggested that the total energy output from the Sun also dropped in these years. This reduced output from the Sun has lead experts to suggest that this caused the river Thames in London to freeze every winter, to such an extent that an open air market and fair could be held on the frozen river.

The theory of the Sun randomly reducing its overall output in the past years has also been linked to large rivers reducing their flow in the years when the solar activity has been low. In other parts of the world other rivers have exhibited the opposite effect, and when solar output returns to normal so do the river levels. Solar activity can play an important role in the radio emissions that the Sun generates. Radio emissions from the Sun come randomly, although when sunspots can be seen this is a good indication that there will be some radio emissions. Larger sizes and numbers of sunspots usually mean more radio emissions, but the Sun has thrown out the odd surprise now and again. Radio emissions from the Sun can come in several different forms and over several different frequencies. Depending on which source of information used the description of each type of radio emission can vary very slightly. The following list has been compiled from many different sources, and common ground has been found from each different description:

1. Quiet Sun: This is taken to mean a low level of solar activity, with little or no sunspots visible, and with X-ray emissions that are below a C class flare. Other radio emissions are continuous across all wavelengths and originate from thermal radiation. For example, the Sun at solar minimum.
2. Type 1: A noise storm composed of many short bursts, from a tenth of a second to fifteen seconds. They are of variable intensity and may last for many hours or days.
3. Type 2: (Slow drifting bursts) These bursts are caused when a shock wave from a large flare travels up through the Sun's atmosphere. The shock wave is caused by material being thrown off by the flare. They are of narrow band emission and slowly sweep from high to low frequencies over several minutes.
4. Type 3: (Fast drifting bursts) These bursts are narrow band and drift rapidly from high to low frequencies over several seconds. They are associated with active regions of the Sun's surface, for example a solar flare or large sunspot.
5. Type 4: Continuum emissions can last from many hours, and are associated with major flare events, beginning soon after the flare has erupted and reaching its maximum intensity.
6. U-burst: Sometimes called Castelli-U, they can last for several seconds and the wavelength changes rapidly, decreasing and increasing again. They are often associated with flares and have a similar origin as Type 3 active regions on the Sun's surface.

Although the above are different types of radio emissions, some occur more frequently than others, and for our purposes the characteristics of each do not need to be known. It takes an experienced person, far more experienced then the author, to tell the difference. It space is an issue two examples of solar activity are shown in the Figs. 7.15 and 7.16. These have been received by the author using the Radio Jove receiver with its dual di-pole antenna and using the program Radio-SkyPipe.

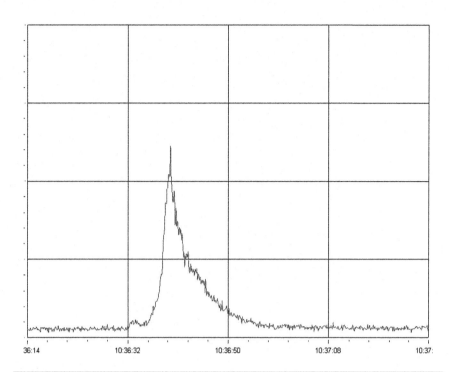

Fig. 7.15 Single Type 3 burst

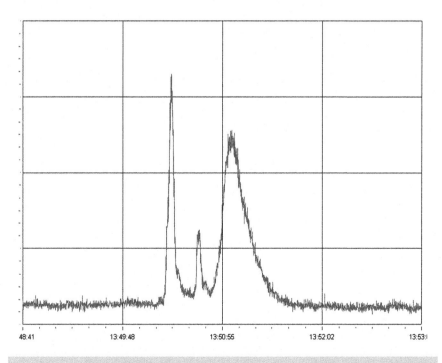

Fig. 7.16 A triple Type 3 burst

This image Fig. 7.15 is a Type 3 burst note the shape is like a shark's fin. Note also the rapid rise (the almost vertical line on the left hand side) and the slow return to background again (the shallow angled right hand side).

The image Fig. 7.16 is of a triple Type 3 burst.

Both images are of one particular type of solar activity. It is highly recommended that one visits the Radio Jove website where there are far more examples of other types of solar activity, along with audio records made of the Sun using the Radio Jove receiver.

7.8 Radio Emissions from the Planet Jupiter

Radio emissions from the planet Jupiter were discovered accidentally in 1955 by B. Burke and K. Franklin of the Carnegie Institution in Washington, DC. The two observers were carrying out experiments using a new type of receiving antenna. This antenna was designed to work at a frequency of 20 megahertz. When the receiver and antenna were switched on they picked up all sorts of signals, both wanted and unwanted. The unwanted parts were thought to come from the usual sources of interference. By the process of elimination the two observers worked their way through each of the unwanted signals and either removed the sources of interference or made allowances for them on the receiver's output chart. But still there was an irritating signal that seemed to come and go with no particular pattern. The two observers thought this random signal may come from the sparks generated by vehicle ignition systems. Then it was noticed that although the interference didn't happen every day it did happen at roughly the same time of day when it did. They made recordings of the interference and tried in vain to track down its source. After they had exhausted all possible sources of interference the two observers then thought the interference was possibility more extraterrestrial than terrestrial. But from where?

Planets at that time weren't even on the list of targets for a radio astronomer. Further studies found that the planet Jupiter was within the beam of the antenna every time the interference was received. So their attention turned to Jupiter, and to their mutual surprise Jupiter was indeed found to be the source of the interference. The question of how the planet produced these radio emissions was the greatest mystery. Due to the sound of the interference it was first thought that it could be lightning discharging high within the Jovian atmosphere. It was an excellent piece of detective work on the part of Burke and Franklin, and full credit to them both. A few years earlier, in Australia, the same type of interference was picked up at the frequency of 18 megahertz, but the connection to the planet Jupiter wasn't found.

Unlike the Sun, which follows the same path through our skies each season of the year, Jupiter can be anywhere in the sky within a short distance north or south of the ecliptic. The ecliptic being the line the Sun traces through the sky over the course of a year. In recent years Jupiter has been in the southern hemisphere, which is great if observing from the southern hemisphere, but for observers in the northern hemisphere this has proven a bit of a challenge for the Radio Jove observer.

7.8 Radio Emissions from the Planet Jupiter

This is because the antennas need to be double the usual height required when Jupiter is in the southern skies, in order to get enough gain from the antenna to receive the radio emissions from Jupiter.

Anyone familiar with using an optical telescope will know through experience that unless the "seeing" is very good it is not worth bothering to attempt to observe objects within approximately 15–20 degrees or so in altitude of the local horizon. This is due to the fact at this altitude we are looking through the greatest thickness of the Earth's atmosphere, and this part of the atmosphere contains all of the light pollution and the pollution from industry, it also contains a lot of dust (especially when an Icelandic volcano erupts) which can produce some nice sunsets. This can be also true with radio astronomy. The radio waves must travel through this greater thickness of atmosphere, and they can also suffer in the same way as optical light, although radio waves suffer to a lesser degree then optical light. So it is a case of waiting for Jupiter to get in to the right position for observation with the Radio Jove receiver, in the same way as with an optical telescope.

When viewing optically the "seeing" is better if the object is at a high altitude as viewed from an observing site, and also when the object is at opposition. At opposition an object is usually at its closest approach to the Earth at the same time as it reaches opposition. This is more or less the same when observing Jupiter with the Radio Jove receiver, although more experienced Radio Jove operators agree that a month or so before the planet reaches opposition is actually the best time to observe using the Radio Jove receiver. Although this may be the ideal time to view, as long as the Sun is below the horizon and the ionization in the ionosphere is at a low level it should be possible to receive radio emissions from Jupiter.

Radio emissions from the planet Jupiter are divided into two types, with the first known as an "L" burst, the "L" standing for the word "long". These L bursts have a sound like gentle waves from the sea washing up on a sandy beach. Their sound has also been described as like a breeze blowing through a leafy tree. The second type of radio emission from the planet Jupiter are the "S" bursts, with the "S" standing for the word "short". These have been likened to the sound that hailstones make raining down onto a corrugated metal sheet or roof. Some refer to the noise of an "S" burst as sounding like the sizzling of a rasher of bacon when it is first placed into a hot frying pan. Both these sounds can be heard separately or together, but they are always heard with the white noise sound which is the sound of galactic background radiation.

If using the Radio-Jupiter Pro program it should be possible to identify which type of radio storm is received. There are three basic types of radio storm, and these are called A, B and C. These originate from different latitudes from within Jupiter itself, although we need not go into too much detail about the origins of these here, as a whole book could be dedicated to this subject. As Jupiter spins on its axis these different zones move through Jupiter's powerful magnetic field generating the radio emissions. Io, the innermost moon of Jupiter, plays a part with the radio emissions from the planet, and this is known as the Io effect. As Io orbits around Jupiter it produces a cone-shaped area in front of the moon itself. This cone is made up of charged particles, and is a type of plasma. This has the effect of creating a beam which concentrates the radio emissions, in the same sort of way that a lighthouse concentrates a beam of light. As Io orbits Jupiter, if the cone or beam of radio emissions points in the direction of the Earth, there is a greater chance of receiving these emissions. These storms are known as Io A, Io B and Io C. Some of these radio storms are more predictable and are of a stronger intensity than others. It is suggested to start with the more powerful storms and then try for the more unpredictable and less powerful ones, having gained some experience in their detection. A good place to start is with Io B and Io C storms, as these can be quite strong in nature. These are made up of mostly "S" burst activities, although this is not cast is stone and "L" bursts may be heard at the same time. The not so powerful storms are the Io A and the non-Io storms. These mostly consist of "L" bursts, but as before it is possible to pick up the odd "S" burst here and there.

The radio storms from Jupiter can be predicted with a certain amount of accuracy. Although it is sometimes found even if Jupiter is in the right part of the sky, in the center of the antenna beam, and Radio-Jupiter Pro is indicating that an Io B storm is taking place, the receiver is picking up absolutely nothing. This doesn't mean there is anything wrong with the equipment or the software. There are many factors that can stop the radio emissions getting from Jupiter to the antenna. The ionosphere may still be partly ionized, especially if solar activity is high. This ionization usually makes itself known by the large number of stations the Radio Jove receiver is picking up. It is also possible that the aurora has decided to put on a show and is playing havoc with the atoms in the upper atmosphere. It may even be something as simple as Jupiter deciding not to go along with the observing plan. Whatever the reason, it just means luck wasn't in that particular night. Be prepared to be surprised by Jupiter now and again. For instance, nothing may be shown on any of the predictions but the receiver is picking up quite a good number of "L" and "S" bursts.

Limitations on space and the large number of different factors involved with these type of radio storms from Jupiter, such as length of duration of each storm, the intensity and strength of each storm, also the varying types of storm, it is strongly advised to visit the Radio Jove website, and once there to follow the links to the Radio Jove archives. At the Radio Jove archives there are lots of excellent examples of Radio-SkyPipe images of both the Sun and of the planet Jupiter. This archive is also a good starting point to familiarize oneself with what to look for in relation to both the Sun and Jupiter. The images may at first seem to look all the same to start with, but soon familiar patterns will be seen forming, especially from the Sun. While at the Radio Jove website audio recordings of the planet Jupiter will be found.

7.9 References for More Information

If only having the remotest interest in receiving the radio emissions from the Sun or the planet Jupiter, a good starting point should be the second edition of the book "Listening to Jupiter" by Richard S. Flagg, published by Radio-sky in 2000 (if a copy can be procured). Flagg's excellent book is dedicated to the study of the Sun

7.9 References for More Information

and Jupiter by radio, covering all aspects of the study of the Sun and Jupiter by amateur radio equipment, not just the Radio Jove receiver. There is a full chapter discussing the Radio Jove project which goes into more detail and there is some mathematics involved. The book is written in a very readable way, with some excellent anecdotes.

The above book and all the computer programs discussed within this chapter, plus many other resources such as books and computer software, are available from Radio-Sky publishing at http://radiosky.com. Even if choosing not to go down the Radio Jove path, it is well worth a visit.

Another excellent website is the Radio Jove Project website at http:/radiojove.gsfc.nasa.gov/. From this site it is possible to find out all about the Radio Jove project, and download newsletters. The newsletters are full of information about any forthcoming events and they include images of Radio Jove receiving sites from all around the world. Some of these images can make one feel quite jealous. It's a good idea to download some of the older newsletters to see how much things have changed. Hear samples of the radio emissions from the Sun and Jupiter, and also hear examples of the galactic background radiation and some of the common sources of interference. Find Radio-SkyPipe images of the different types of radio emissions from the Sun and Jupiter, and those interested can download and print the order form for the Radio Jove kit from here. There are links to other websites of interest.

Another useful website is http://www.swpc.noaa.gov/ftpdir/indices/events/events.txt. This site produces a list in simple text format that relates to the solar activity for that particular day. If one thinks a solar flare or other radio emission has been captured, simply check this site and the particulars of the flare can be identified. Take a good look around this site as there is lots of information and a glossary of terms used in solar radio astronomy.

Chapter 8

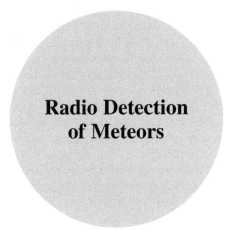

Radio Detection of Meteors

8.1 Meteoroid, Meteor, Meteorite and Micrometeorites

Most meteors that can be seen on a clear night, especially on the nights of a meteor shower, range in size from a grain of sand to the size of a pea. These meteors come from the debris left over from the tails of comets that have orbited around the Sun, in most cases many years ago. Mixed in with this cometary remains are small particles of debris that have come from other sources, such as the collision of two or more asteroids within the asteroid belt or the rubble left over after the solar system came into being. These small particles are drawn towards the inner solar system by the Sun's gravity, they can get trapped within the debris left by comets. These small particles of cometary and other debris are known as meteoroids and have been left floating in space for many years after the comet has passed. They are gravitationally attracted to each other and build up to form dense regions, called meteor streams.

As the Earth makes its annual orbit around the Sun, it encounters the odd bit of debris. As a result we will get the odd sporadic meteor. It is possible to see approximately 10 sporadic meteor trails every clear night. If the Earth encounters one or more of these densely packed meteor streams which are full of small fragments or meteoroids. There can be a large number of meteoroids entering the atmosphere in a short time, even over the space of a few nights, and this is what is termed "a meteor shower". There are several well-known annual meteor showers. As these meteoroids enter the Earth's atmosphere their name changes and they become meteors. These meteors are travelling at high velocities, anything between 8 and 85 kilometers per second (5–50 miles per second), they burn up in the atmosphere and leave a streak of light in the sky.

From the Earth's surface all the meteors on the nights of a meteor shower seem to originate from a single point in the sky, and this point is called the radiant.

Meteor showers are named not after the comet which produced the debris left floating in space, but after the constellation that contains the radiant. For example, the Leonid meteor shower. The radiant for the Leonid meteor shower is the sickle shape in the constellation of Leo the lion, this point is sometimes referred to as the backward facing question mark. Although there isn't a meteor shower every night, it is estimated that the Earth picks up approximately 20 tons of debris from space every day. This debris comes from micrometeorites, these are microscopic dust particles that instead of burning up in the atmosphere are quickly slowed down by it and, because of their very small size, they just float down from the sky unnoticed, in the same way as sand or volcanic ash but not in the same localized volumes as sand and volcanic ash. There are of course larger pieces that fall from the skies. These are not linked to cometary debris, they can be anything from a stray piece of the asteroid belt to pieces of the planet Mars.

Not all meteors are made of stone. Some are made of iron and nickel or a mixture of the two metals. There are also stony iron meteors. When a meteor hits the Earth's surface it is no longer called a meteor, it becomes a meteorite. Since most of the Earth is covered in water there is a good chance that any meteorite will be lost in the Earth's oceans. If a meteorite hits the ground it can cause a crater such as the meteor crater in Arizona in the United States. This was of the iron nickel type, as samples of the original meteorite have been found within the ejector that surrounds this huge crater. It is estimated that the meteorite that produced this crater was around 45 meters (148 feet) in diameter, but that most of the original meteorite vaporized on impact. Since the invention of satellite imaging other large craters have been found in different parts of the world, because of the weathering caused by the Earth's atmosphere these other craters went unnoticed until they were imaged from Earth orbit. A large meteorite impact is thought to have brought about the extinction of the dinosaurs.

Meteorites can be found in most of the deserts around the world, as they are less likely to be disturbed by human activity. Most deserts have very low annual rainfall, so any meteorite will not have suffered from too much weathering before it is found.

An excellent place to find meteorites is Antarctica. As meteorites travel through the Earth's atmosphere they get very hot from the friction developed by the atmosphere acting as a brake to slow them down. The outer surface becomes dark in color, almost a matt black in appearance, as the heating from the atmosphere chars the meteorite's surface. With everything in Antarctica being covered in ice the landscape is totally white, with the meteorites being black the contrast between the two makes them very easy to see on the ground. The extremely cold temperatures and the dry atmosphere of Antarctica help to preserve the meteorites. NASA also funds regular expeditions to Antarctica to search for meteorites, these have been very successful.

Theories suggest that life travelled to Earth from elsewhere, such as the planet Mars, by hitching a ride on a meteorite or a comet. The term "life" loosely refers in this context to something along the lines of bacterial life, or single celled organisms, or just the right chemical cocktail of organic compounds and amino acids, not life forms which we would be familiar with in everyday life. This theory is called Panspermia, but there are quite a few flaws within this theory. How does life get off the planet Mars in the first place? How does life survive the cold and radiation of space if it's clinging to the outside of a meteorite? How will life survive the high temperature when the meteorite is entering the Earth's atmosphere? Theories suggest that life could survive if it was below the surface of the meteorite, but how did it manage to get below the surface of the meteorite? And the big question: if life on Earth originated from Mars, where did life on Mars originate? Life travelling to the Earth on board a comet, protected from the hostile environment of space within the cometary nucleus is a more plausible theory, but this is a discussion for another day. The meteors that we are concerned with in radio detection are the small sized meteors that burn up in the Earth's atmosphere at an altitude of approximately 80–100 kilometers (49–62 miles).

8.2 What Is Radio Detection of Meteors?

The first radio observations of meteors were carried out by J.S. Hey in the mid to late 1940s just after World War Two. Hey and his colleagues, working with army radar after the Second World War, were following up on reports from radar operators about the existence of short duration echoes that they reported hearing. A number of theories were put forward to explain what these unwanted echoes might be, from attempts at jamming incoming radar signals by the enemy, to cosmic rays from outer space, to the notion that the echoes were in some way connected with meteors entering high up in the Earth's atmosphere. Hey and his colleagues proved that these short-lived echoes did in fact come from meteors entering the Earth's atmosphere.

Hey found that the reflection of the radar signal was due mostly to the fact that as the meteor enters the Earth's atmosphere it leaves a trail behind it, a meteor trail. This trail is due to the localized heating and ionization of the atoms in the atmosphere. Hey and his colleagues found that the reflective qualities of this ionization within the meteor trail had the same reflective properties on the radar signal as if it had been reflected off of a metal object. Therefore, an ionized meteor trial was effectively like having a large copper wire in the sky, if only for a few seconds. Hey found that the number of meteor echoes were considerably higher at certain times of the year, on the days of meteor showers. Hey and his colleagues were also the first to observe a meteor shower during daylight hours using radar equipment. Hey theorized it would be possible to use the times of high meteor activity to conduct experiments into the structure of the upper atmosphere.

The practical theory behind radio detection of meteors is really quite easy to understand, but some of the descriptions of the process for undertaking this detection have been grossly over-simplified and some who have tried it have been disappointed with the results, despite their sincere efforts. There are two ways that meteors can be detected entering the Earth's atmosphere using radio equipment, one is directly from the meteor itself, if it is large enough. The second and more practical way is by the trail that the meteor leaves in the sky, as this offers a far larger target. As mentioned above, meteors entering the Earth's atmosphere produce a streak of light, and as they travel leave a meteor trail across the sky. These trails can last from just under a second to nearly a minute, depending on the size and angle at which the meteor enters the Earth's atmosphere. These trails are caused by the friction of the meteors entering the Earth's atmosphere at very high velocities. The heat generated by the friction heats the atoms within the atmosphere and causes them to become ionized. In an ionized state the atoms in the atmosphere become highly reflective to radio waves, and can reflect or bounce a radio signal off this ionized trail, or directly from the meteor itself, listen for the return echo.

This is how radar works; the radar equipment sends out a signal and waits for the echo to return. The signal travels at the speed of light, 300,000 kilometers per second (186,200 miles per second). The radar operator knows the speed of the outgoing signal, and if they also know how long it takes for the return echo to get back, the operator can work out the distance to an object, such as an airplane. If the radar signal is monitored, any change in the time it takes for the echo to return tells the radar operator whether the airplane is flying towards or away from them. Unfortunately radar systems are quite expensive to purchase, not to mention the high costs running and maintaining of such a system. So we need to borrow a radio signal to bounce it off the meteor's trail in order to detect them. In the past it was quite easy to do, by just borrowing the carrier wave signal from a television transmitter or a radio station transmitter.

The transmitter needs to be situated ideally below the local horizon and not directly detectable. The receiver is then set to the carrier wave frequency of this transmitter (ideally a little off the exact frequency to allow for the effect of the Doppler shift on the returning signals, more about this later). The theory is, the carrier wave from the transmitter would be transmitted in every direction including straight up into the sky, but because this transmitter was below the local horizon so that it's signal couldn't be picked up directly, because the Earth was in the way. As a meteor entered high up in the Earth's atmosphere and produced an ionized trail, the carrier wave signal would hit the ionized trial or the meteor itself and for a brief moment would be reflected back down to the Earth's surface, and hopefully in the direction of the antenna and receiver. For a short time a signal would be received, until the ionization in the meteor trail was lost or the meteor burnt up. Please see the Fig. 8.1 for an example. This is a simplified drawing of how radio meteor detection works. It should be noted how the transmitter and receiving antenna are not in direct line with each other because of the curvature of the Earth.

This was the case in the past, but nowadays the old analog signal transmitters are being switched off and replaced by the more modern digital signals. Digital signals can carry more information and take up less space or bandwidth then the older analog signals. So, lots more television and radio stations can be broadcast in the same amount of bandwidth as the old analog signal. There are still some analog transmitters broadcasting, one will need to check there location and suitability to their own observing site, as changes are being made almost daily, and what may be analog now may have switched to digital next week. The analog transmitters work the best, as the way the new digital signal is transmitted and received makes it unsuitable for the purpose that we wish to use it for.

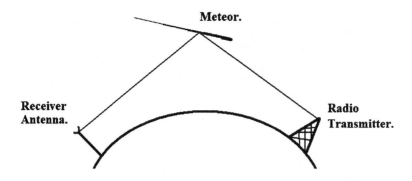

Fig. 8.1 Image showing the theory of radio meteor detection (not to scale)

Luckily for the radio meteor observer there are other transmitters that can be used. There is a powerful space radar in the United States that monitors satellites and other objects known as space junk in Earth orbit. Radio meteor observers in the United States use the signal from this space radar for meteor detection. There are also transmitters in Belgium and France that are used by radio observers in Europe. The French space radar is quite a powerful one to use if living in Europe, and transmits at a frequency of 144 megahertz which is ideally suited for radio meteor detection. Look at the end of this section on radio meteor detection and find a number of websites that could help one find a suitable transmitter to use from their location. Ideally find a transmitter that transmits at a frequency within the region of 50–200 megahertz, as this range is highly suitable for radio meteor detection. If a transmitter within this range cannot be found, as long as the receiver is capable of receiving the frequency from the transmitter then it is worth giving it a try.

If two suitable transmitters can be found it can be of great advantage, especially if the transmitters are in opposite directions from the observers location i.e. north/south or east/west. This can be useful if one transmitter has been shut down for maintenance, also the transmitter that best suits the direction of the meteor radiant can be selected, or even use both at the same time, although this would require more equipment.

Radio detection of meteors offers several advantages over visual observations, as radio observations are not affected by the light from the Moon and therefore light pollution won't be a problem either. There are times when the maximum level of activity for a particular meteor shower is forecast to happen during the day, so even if it were clear it would be too light to observe any meteors. However, radio observations can still be made during the day. Radio observations are also unaffected by the dreaded clouds.

8.3 When to Listen

Although it is possible to hear the echoes from meteors all year round, there are certain days and nights when the number of meteor echoes can increase greatly, these are the annual meteor showers. On these particular days and nights it may be possible to pick up the echoes from several hundred meteors per hour over a short period.

The table below gives some of the better known and most reliable showers, with a brief description of what to expect. Please note the zenith hourly rates on the table are an indication of the expected rates and are an average taken from many years of study by dedicated meteor observers, both visual and radio. These can vary wildly and are not set in stone. It's possible to have half an hour of really high activity and nothing more than a few meteor echoes per hour for the rest of the night. If wishing to investigate further there are other less well known meteor showers with maximum zenith hourly rates of around three per hour. Note that all of the dates used within the table can vary a day or so due to natural variations in the width of the debris path and the Earth's orbit.

Shower name and radiant	Month	Date of maximum	Approx zenith hourly rates	Description
Quadratics	January	Around the 4th	50–100	Narrow maximum
Lyrids	April	Around the 21st	10–15	Moderate display, but can put on a surprise show
η Aquarids	May	Around the 4th	20–50	Not easily seen from high northern hemisphere latitudes
				Best seen in southern hemisphere
δ Aquarids	July	Around the 28th	15–25	This shower has a double radiant
Perseids	August	The glorious 12th	50–80	Rich annual shower, persistent trails
Orionids	October	Around the 21st	20–30	Fast moving persistent trails
South Taurus	November	Around the 3rd	10–15	Slow moving meteors
Leonids	November	Around the 16th	15<	Very fast meteors persistent trails. Can put on a good show from time to time. Is thought to follow a 33 year cycle of activity. Due again 2032?
Geminids	December	Around the 13th	50–100	Richest annual shower, usually an excellent display. Bright trails.
Ursids	December	Around the 22nd	10–20	Can sometimes put on a good show

Any good monthly astronomy publication will have details of forthcoming meteor showers, or check online using any of several good astronomy websites. Usually the dates given in astronomy publications can give the duration for a particular meteor shower anything up to 2 weeks. Along with this information will be dates and times where the maximum activity is expected to be. This expected date and time of maximum activity is the best estimate with the information available at the time of publication. Take this date and time of maximum activity, and allow 24 hours either side of the forecasted maximum, this will give the best chance of hitting the point of maximum activity. Depending on the position of the meteor radiant, the general rule is that meteors are more common, both visually and by radio detection, after midnight local time. This is because the Earth's rotation at this time has started to turn the Earth directly into the stream of debris, as seen from the local location. This can be thought of in these terms, "you only get squashed bugs on the windshield of a car never the rear window." So the best times are the hours from midnight to 6 am local time.

There are limitations with radio detection of meteors, however, just as there are observing meteor showers visually. Some meteor showers will be picked up better than others, depending on the height of the radiant as seen from the observing site. This is why it is a good idea to have more than one transmitter to tune into. It can be a case of trial and error but the effort is worth it.

8.4 What Equipment to Use

One may have already heard that any radio receiver can be used to listen for meteor echoes, and this is true to some extent, as long as the receiver can be tuned accurately to the right frequency of the signal being used to reflect off the ionized meteor trail.

A receiver with a good frequency range and the ability to tune it accurately is a must. Ideally use a receiver with a digital display, as these can be tuned far more accurately than a receiver with a sliding scale. Digital displays are easier to read, which can be handy if using them at night. A receiver with the ability to have an external antenna fitted is an advantage. It is possible to detect meteor echoes using a whip antenna, but it is far better with an external antenna.

Once the receiver has been tuned to the chosen station the "hiss" of white noise will be heard, when a meteor enters the atmosphere and the signal from the transmitter is bounced off the meteor or its ionized trail, in the direction of the receiver, it will be possible to hear a few seconds of music or speech – whatever the transmitter is broadcasting at that moment. The music or speech will fade when the ionization is lost and the signal is no longer being bounced to the receiver. This is fine for proving the theory, but it can make it difficult to follow, because it is not known whether music or speech is going to be heard. It makes any potential meteor echoes difficult to record, as sometimes the speech can be a bit garbled.

A better way by far is to have a receiver that has the ability to receive "upper side band" (USB) or "lower side band" (LSB). In this mode the receiver will inject a tone every time a signal is received. Thus, every time a meteor echo is detected a "Phwoooo" sound will be heard. An example of this sound, is the sound of a sonar ping. As this makes meteor echoes easier both to listen to and record, because one is listening for roughly the same tone each time. It also makes recording the sound easier as the problem of trying to set the recording levels for the different types of music and speech that is broadcast, doesn't arise. At first each "Phwoooo" sound may all sound the same, but with practice it will be noticed that each one is slightly different in character. This is because meteors enter the atmosphere at different angles, travel in different directions and at different velocities.

The sound of the echo that is received will be Doppler shifted, the pitch of the tone of the echo can and does change randomly. Think of those whistles that have a plunger in the base, which are usually given to children, the plunger when moved changes the length of the whistle's internal dimensions, allowing the air to escape at different places. Blowing the whistle and moving the plunger up and down changes the pitch of the tone. This is the sort of sound that will be heard.

A good receiver suitable for the task of meteor detection is sometimes referred to as a radio scanner. These scanners can be quite small in size, no bigger than a conventional pocket radio or walky-talky, unlike a normal radio that can only pick up FM and AM stations, scanners can pick up almost every type of commercial radio frequency, and in a number of different modes including upper and lower side bands.

The cost of these scanners when purchased new can vary quite widely, and they can get quite expensive depending on the number of tasks they can perform. If just wanting a scanner to use for meteor detection they can be purchased second-hand off internet auction sites or from radio equipment swap shops quite reasonably.

If choosing to purchase a radio scanner second-hand, make sure it has the ability to be used with a mains/grid power supply, and that the mains/grid power cord or transformer is included in the purchase. This is because the scanner will remain switched on for a number of hours, and save time and money not having to buy and fit replacement batteries.

Also, make sure there is an audio output socket for headphones or an ear piece somewhere on the scanner. It really doesn't matter if this output is stereo or mono as an adaptor can be purchased, which can be used to link both channels of a stereo input of a pair of headphones together, so sound will come out of both sides of the headphones and not just the one, if the scanners output is a mono. This is not true stereo, but it is perfectly adequate for our purposes. This will be needed to connect the scanner to a computer in order to record the scanner's output onto the computer using its sound card.

A good entry-level radio scanner for radio meteor detection is the Yupiteru range of radio scanner, model numbers such as MVT-5000, MVT-7000, MVT-7100 and the MVT-7200. Some of the earlier scanners are getting quite old now and are being replaced by newer models, but the older scanners are ideal for detecting meteors.

Yupiteru radio scanners are very sensitive and have enjoyed a good following with radio enthusiasts who hold these scanners in high regard. The Fig. 8.2 is of a radio scanner that can be used to detect meteor echoes.

One slight drawback with the Yupiteru scanners is the instruction manual. The original Yupiteru instruction manual is notoriously difficult to understand, due to the translation from the original Japanese to the English version. An instruction manual is needed to understand the original instruction manual. Several brave people have tried writing their own versions of it, they have made an excellent job

8.4 What Equipment to Use

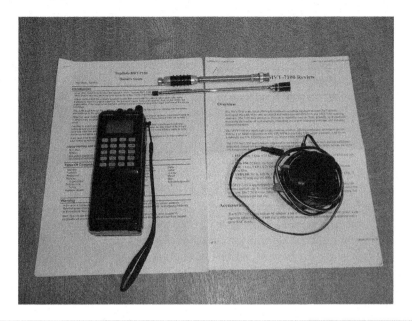

Fig. 8.2 An entry-level radio scanner suitable for radio meteor detection

of doing so. Entering the name and model number of a Yupiteru scanner into a search engine can soon track down one of these rewritten manuals and a copy can be printed. These scanners can perform a vast range of tasks, if only using the scanner for radio meteor detection there is no need to know how everything works, just the basics and how to set up the scanner for the chosen frequency, and store this frequency in the scanners memory. It has a capability to store 1,000 different channels, so no need to worry about filling its memory.

This is just one make of radio scanner and don't be swayed either way into purchasing a Yupiteru scanner. There are lots of other scanners on the market, and some may be far better than the Yupiteru scanners. There are some scanners that can monitor a number of channels simultaneously, with computer systems to match, it would be unfair and unwise to make comments about them having no experience in using them.

These types of scanners and computers, will of course come with a hefty price tag and it would take a serious radio meteor observer to own such a setup. The radio scanner mentioned above and other scanners like it are intended to be an entry level to the hobby with minimum outlay to the user, and therefore must be considered as the minimum equipment needed to feasibly detect meteors by radio. In fact this radio meteor detection system has been operated for a number of years now and the scanner and the home-made antenna (this will be covered next) work well together, and as long as the system continues to work there seems no reason to change to more advanced and expensive equipment.

Antenna Options

Radio antennas come in many different forms. There are lots of different factors that need to be taken into consideration when antennas are designed. These include the wavelength at which the antenna is to operate, the polarization of the incoming signal, and the gain needed from the antenna. Some antennas are highly directional, like multi-element Yagi's, or hardly directional at all, as in the case of a single di-pole antenna. To give an example of the concept of directionality in antennas, we can use an optical telescope and different eyepieces. When observing a planet through an optical telescope, the planet's angular measurement is small, so a short focal length eyepiece is used to give a high magnification so detail on the planet can be seen. By using a high-powered eyepiece we trade field of view for magnification. Think of this as a highly directional antenna, we are trading the field of view, or "pick up", of the antenna for maximum gain in one direction.

If we change the eyepiece of our optical telescope to one of a much longer focal length and now observe the same planet, the planet will be a very small disc in the center of the eyepiece, we will have a good low-power view of the sky around the planet. We are now trading magnification for a greater field of view. This would be the principle of the field of view or pick up of the single di-pole antenna. This offers a greater area of detection from the antenna, but we lose out on gain in any particular direction. Obviously an antenna will cover more area of the sky then an eyepiece of a telescope, but this should demonstrate the principle. We therefore need to choose the right design for the job we want the antenna to do.

If we choose an antenna that is highly directional we can only observe that part of the sky which the antenna is pointing at, so meteors that appear in other parts of the sky will be missed. If we choose an antenna that has very little directionality associated with it, like a single di-pole antenna, we cover quite a large area of the sky, but the gain in any particular direction won't be as great as a highly-directional antenna. So, we have to choose between gain and directionality of an antenna, depending on the observing site and proximity to the radio transmitter available for use. As a general rule use the antenna that has the least directional quality that will work. If getting meteor echoes with a signal di-pole, great; if not, try and find another transmitter to use. If this doesn't work, try a two element Yagi, and if this doesn't produce any echoes try and find another transmitter. Again, if this doesn't work then try a three element Yagi, but don't go any more then this as this is just narrowing the field of view too much. Try moving to a different observing site, one that gives a better view in the direction of the transmitter. There is a lot of experimentation needed with radio meteor detection, and once everything is working it is understandable if one is loath to change it.

Meteor echoes come from a height of approximately 70–80 kilometers (43–49 miles) in altitude, but the height at which they will be able to be received above the local horizon is dependent on how far away the transmitter is located. The further away the transmitter is, the lower in the sky that the echoes will be able to be picked up. A little experimenting will be needed. The height at which the antenna is placed isn't crucial. It was found that 2 meters (72 inches) works very well. In fact, the antenna was once tried at 4 meters (144 inches) and it didn't work as well. In relation to the polarization of the antenna, it is worth knowing that as a rule most manmade signals are vertically polarized. But there are bound to be exceptions to this rule. If in doubt, position the antenna so that it is vertically polarized.

In radio meteor detection we are receiving a reflected radio signal from the meteor itself, or more likely from its ionized trial, so the polarization of the original signal can be changed after its encounter with this ionized trial. For this reason some radio meteor observers use a cross di-pole antenna. This type of antenna has a vertical element and a horizontal element to it. The theory behind this type of antenna is whatever the polarization of the incoming signal is, whether it is vertical, horizontal or even circular in nature, the antenna should pick it up. Once a suitable transmitter as been found, it is possible to construct a simple half wave di-pole antenna, as described within the section on antenna construction in the chapter on the Radio Jove receiver.

The frequency of the transmitted signal will need to be known, in order to cut the antenna to the correct length of half the length of the signals wavelength. To convert frequency to wavelength, and vice versa, please refer to the chapter on "basic physics". An antenna made of copper wire of the same gauge as used for the Radio Jove antennas will be fine. An antenna cable of 75 ohm resistance should be used, such as the RG59/U used in the construction of the Radio Jove antennas, as this is an excellent match for use with a di-pole antenna. The antenna wire can be fixed to something none conductive, such as a wooden garden cane, using cable ties or electrical tape to support the wire, so that it can be easily mounted. Another idea is to use an old fiberglass fishing rod for this purpose. After all the hoops and other fittings have been removed, the antenna wire can be cable tied to the fiberglass fishing rod and mounted to the side of an observatory of other out building. This has a couple of advantages over wood. Fiberglass will not rot, it is very strong and light weight, and is a good electrical insulator.

A connection will need fitting to the other end of the antenna cable in order to couple the antenna to the scanner, and this will probably be a BNC connector as described in the chapter about the SuperSID monitor. Make sure the BNC connector is also of 75 ohm as there are 50 ohm connectors available, if a 50 ohm connector is fitted this will produce an impedance miss-match within the antenna and cable causing the antenna performance to be reduced. There will probably be an impedance miss-matching between the antenna and the scanner itself anyway, but nothing can be done about this. However anymore miss-matching can be stopped, by fitting the correct connections and using the correct antenna cable. The length of antenna cable used to carry the incoming signal from the antenna to the scanner should be kept to a minimum, but as with the SuperSID antenna cable do not put the cable under undue stress by pulling it too tight. If a single di-pole doesn't give enough gain to receive meteor echoes, by all means try a two or three element Yagi antenna, these may be made or purchased commercially from antenna suppliers. A good book with lots of information on the theory of antennas and antenna suppliers is "The ARRL Antenna book" published by the Amer Radio Radio League.

The 22nd edition of the ARRL antenna book contains nearly 1,000 pages covering just about everything there is to know about antennas, including how they work, antenna patterns, antenna construction and antenna suppliers, plus useful websites. It is highly recommended.

8.5 Software for Recording and for Analyzing

Once the radio meteor detection equipment is working, the next logical step would be to record the output of the scanner. Recording the output from the scanner has the advantage of giving a permanent record of a day or night's observing. Equipment can be left recording while in bed or at work, the recording can be processed at a more convenient time. There are two programs that are good for this purpose, they have already been mentioned within the chapter on the INSPIRE project. These programs are Spectran and Spectrum Lab, as these are highly suited to the hobby of radio meteor detection.

The first thing is to connect the output from the scanner to a computer. This can be done by using a cable fitted from the headphone socket on the scanner to the inline connection of the computer. These cables have a male jack plug fitting on both ends, and can either be mono or stereo. Either will work, but if a stereo cable is used in a mono output, a small adaptor can be purchased to link the mono channel to both the stereo channels to give the effect of a stereo output. As with other projects covered here, the microphone input must be muted by using the audio property settings on the computer. If the computer doesn't have an inline input the microphone input can be used, but keep a close eye on the input levels when setting the input volume. This is to ensure that the inputted signal, doesn't overwhelm the computer's sound card.

Both programs are equally good, but Spectran is easier to use when first starting out.

Spectrum Lab can be programmed to count the number of meteor echoes, but this task can be a little fiddly and should only be attempted when familiar with the full workings of the program. It would be advised to try both programs and see which program does the job best. If one has an old laptop that is getting on a bit and rather slow, this aging laptop can be left in an observatory overnight and used to record meteor echoes, without the worry of leaving a more expensive computer outside in an observatory, where it may suffer damage from damp in the air. In this case Spectran is the preferred recording software as this seems to work better on slower machines.

It is possible to use the basic recording software supplied with the computer, if it has any, but if using one of the programs mentioned above advantage can be taken of the program's on screen facilities. The feature that is most useful and interesting to use on these programs, apart from their ability to record, is the spectrogram. This looks like a "waterfall" effect and begins as soon as the program is started. The bottom window on the screen gives a visual representation of the incoming signal in a waterfall effect flowing from top to bottom. This is extremely useful when listening for meteor echoes, as an echo is received a visual representation of this echo will be seen on the screen. Some echoes can be quite faint, but the computer is far more sensitive, and if unsure whether an echo was heard, just look at the screen to see the echo registered within the spectrogram window.

Within the spectrogram window it is possible to watch the frequency of the incoming echo drift, either side of the carrier wave frequency as the incoming echo is slightly Doppler shifted, from the transmitted frequency of the carrier wave. Depending on the size of the meteor and its angle of entry into the Earth's atmosphere it's possible to receive a number of different echoes. This is due to the amount of radio signal that is bounced off the meteor or its trail to the scanner. Double echoes can be received as well. This can be two meteors entering the Earth's atmosphere at the same time, or one echo that has come from the meteor itself and the other from its ionized trail. This can sometimes be noticed if one echo is much shorter than the other. The short echo has probably come from the meteor itself. This can soon disappear as the meteor burns up in the atmosphere and becomes too small to reflect any radio waves. The longer echo will probably be the ionized trail left by the meteor, as this will continue to reflect radio waves until the ionization has been lost.

While watching the spectrogram display, it can be observed that a stronger echo will produce a thicker line on the display than a weaker echo. This can be useful to know in case a really strong single echo is received followed by a number of weaker echoes. This appears as one echo that changes to a number of echoes on the spectrogram display. This could be caused by a larger meteor entering the Earth's atmosphere which produces the single echo, then as the meteor starts to fragment into smaller pieces as it burns up in the atmosphere this could produce multiple echoes, as each of the meteor fragments reflects the radio signal.

As mentioned in earlier chapters, these programs have a "de hum" filter that can be set at 50 hertz or 60 hertz, and this is useful for removing "mains hum" from the incoming signal. These should be used if the scanner is picking up mains hum. Try and keep the scanner a distance away from the computer, this need only be in some cases 1 meter (39 inches) or the length of the connecting cable used to connect the scanner to the computer, doing this will help stop the scanner from picking up interference from the computer itself. A word of caution, not all computers produce the same level of interference; some are far worse than others. One has an old laptop which produces much less interference to the radio scanner than a more modern laptop a friend has. This is why the old "faithful" laptop is hung on to, if only for radio meteor observing. The reason why the old laptop is less of a problem is not clear, it maybe to do with the processing speed of the computer being at a lower frequency, or just something as simple as a lower noise level being generated from the electronics within the computer itself.

Both the above mentioned programs can be set to record quite easily, and the recorded sound file will be automatically saved before exiting the program. See the chapter on the INSPIRE receiver for more information about this. Both programs can be used for playing back the recorded sound file later.

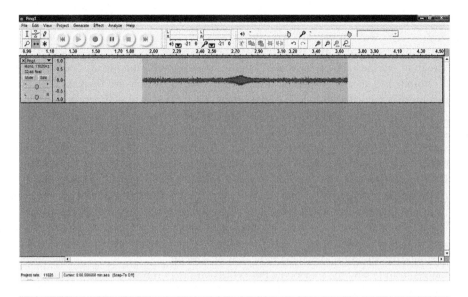

Fig. 8.3 Screen shot of the Audacity analyzing software. With meteor echo

Once the recording has been made of a night's or day's worth of meteor echoes, it's a good idea to save the meteor echoes and discard the parts of the recording with no meteor echoes. Doing this can condense several hours worth of recording into a few minutes, containing the best meteor echoes from the hours worth of recording. This can then be saved to any format the user wishes (avoiding MP3 formats if possible).

A good way of doing this is to use editing software. There is a program that seems to be ideally suited for the purpose of analyzing meteor echoes, and that is available for downloading from the internet, this program is called "Audacity".

Audacity was developed and written by a team of volunteers from all around the world, and Audacity is available for download free of charge for private use, but as always please read the terms and conditions of its use. This software is not over-complicated and is very intuitive and easy to use. Looking at Fig. 8.3, a screenshot of the Audacity program, it is possible to see some of the main operations of the program. Top left is the play, pause, fast forward and rewind controls. In the center are simple cut and paste tools. If unsure of any of the program's functions click on the "Help" tab at the top left, and then click on contents. This then reveals a full list of functions and instructions on how to operate the program.

Looking closely at the screenshot, below the tool bar at the top, a graphic representation of a meteor echo recording can be seen. In the center of this graph, a widening of the graph can be seen, this is a meteor echo, it lasted for about a second but it can easily be seen. Once familiar in spotting them they can be easily seen, even though they vary in length and strength. It is possible to receive some unwanted signals and/or interference with the meteor echoes. These unwanted signals will have to be removed from the recording and discarded using the "cutting tool".

A good feature of Audacity is that, if a potential meteor echo is seen, a slider can be dragged from the left-hand side, and placed on the graph a few seconds before the suspected echo; then press the play button on the control panel above. This will play the sound file from this point so that it can be seen if it is a meteor echo, or interference. This is very useful while learning what to look for within the graph.

When a number of meteor echoes have been collected they can be very easily "cut and pasted" together to form a shortened version of the original recording. A tip: when doing this, allow a few seconds either side of each meteor echo when editing, 2 or 3 seconds will be enough, as if they are placed too close together, when played back it will sound something like a firecracker. Any edited recording can be exported as a WAV file and saved as a separate sound file. The program is also capable of exporting sound files in MP3 format, but this requires the further download of an application. Beware however, this application was tried and it changed some of the computer's settings, so it was quickly deleted. This may have been the computer's operating system throwing a bit of a strop, but it is worth mentioning just in case the same thing happens.

It is possible to import MP3 sound files without this further application. The edited recording can now be played back through either Spectrum Lab or Spectran, observe how each of the echoes causes the frequency to drift higher or lower on the spectrogram window. If a really interesting effect is seen, like a multiple signal echo or a particularly strong echo, both Spectrum Lab and Spectran have the ability to instantly save the display, using this feature it is possible to build up quite a selection of different meteor echo effects.

The display of the sound file on the screen of Audacity can be easily changed, zoom in or out to give a better and more detailed view of the sound file. This is a good feature when working with a sound file of many hours.

With the constant improvements and changes made to free software such as Audacity it would be wrong to give a reference to a particular website from which to download the software, simply type the words: "Audacity download" into a search engine which will provide a number of links to different websites, from which to download the latest version of the software. Just be careful when choosing a source that it appears reputable.

8.6 References for More Information

A good first step would be to listen to the podcast by Steve Carter for the 365 days of astronomy project. Steve Carter wrote and recorded a great podcast on this very subject. Hear some excellent samples of the haunting noises made by meteor echoes that he has recorded, he gives a brief description of how to do it. Even if there is only the smallest interest in trying this project, it is highly recommended to listen to this podcast. It is available from www.365dayofastronomy.org. The date of the podcast is 2nd February 2011. It is titled "Meteor detection by radar for the amateur observer".

There are a number of useful websites where advice on the hobby of radio meteor detection can be found. A quick search on the internet by typing: "radio detection of meteors" into a search engine will be enough to find a list of websites that will give practical advice on scanners/receivers, antennas and the transmitters available for use from one's location. Some sites even give samples of meteor echoes.

Here is a short list of them:

Radio Meteor detection, www.skyscan.ca/radio_meteor_detection.htm
Meteor detection-NLO Radio astronomy, www.meteorscan.com/
American meteor society, www.amsmeteors.org/ams-programs/radio-observing/
Detecting meteors using radio, www.meteorwatch.org/science
Radio meteor listening, www.spaceweather.com/glossary/forwardscatter.html

As a member of the United Kingdom-based Society of Popular Astronomy (SPA), there is an advisory service that members can use to help with finding suitable transmitters and equipment. When using this service include membership number in all correspondence. www.popastro.com.

The British Astronomical Association (BAA) has also just recently started a radio astronomy section. www.britastro.org.

United Kingdom Radio Astronomer Association (UKRAA) www.ukraa.com.

The Society For Amateur Radio Astronomers (SARA) based in the United States, should be able to help with any enquiries concerning radio meteor detection. www.radio-astronomy.org.

Books

"Amateur Radio Astronomy" by John Fielding, published by the Radio Society of Britain.

This book has a chapter on radio detection of meteors. It describes in more detail how the signal is reflected and received, and contains an interesting design for a corner reflecting antenna.

"Field Guide to Meteors and Meteorites" by O. Richard Norton and Lawrence Chitwood, published by Springer.

This book has been very well thought out and put together. It covers meteors and meteorites in great detail and discusses the different types of meteorites, both stone and metallic types and the physical properties of each. It is written in a highly readable way and anyone remotely interested in these "visitors" from outer space would do well to get a copy.

8.6 References for More Information

"Radio Astronomy Projects" by William Lonc, published by Radio-sky Publishers. Covers this subject in a number of places within the book. This book will be quite heavy going for someone first starting out with no experience, but one will grow into it as one's knowledge of the subject increases.

"Meteorites" by Robert Hutchison and Andrew Graham, published by Sterling.

A short book of 60 pages, full of information about the different types of meteorites that have been found throughout the world. It contains some interesting colored images of the crystalline structures of meteorites and describes how observing them with polarized light can show up different chemical elements within the thin cross-sections of meteorites. A good starting point for the study of meteorites. This book is written in a non-technical manner and makes good bedtime reading.

Chapter 9

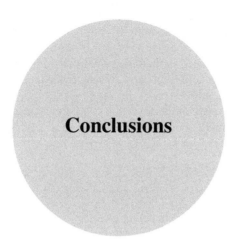

Conclusions

It is hoped that this book dispels the idea that huge parabolic dish antennas are needed in the garden in order to follow the fascinating hobby of radio astronomy. In fact it can be done with quite modest equipment, that won't come with a hefty price tag and that doesn't require a degree in either mathematics or electronics to understand how to build and operate the radio receivers, monitors and scanners. Radio astronomy can be as detailed and in-depth as participants wish to make it. All the projects covered within this book are capable of being built by anyone who can follow an instruction manual and knows the hot end of a soldering iron from the handle.

Most amateur optical astronomers that one has met over the years have come up with some truly ingenious ways of fastening all manner of different accessories to their telescopes, and can talk for hours on end about the thermal efficiency of the chip that is in their CCD camera. Therefore there seems no reason why some of these ingenious ideas can't spill over into the hobby of radio astronomy.

Each of the four projects covered within this book have their pros and cons, and not all will be suitable for everyone for whatever reason, such as lack of space or the problem of interference. Don't let the threat of interference put anyone off trying any of these projects. If interference is thought of as light pollution, one has a good idea of the local light pollution problem and can make an informed judgment regarding what size and type of optical telescope to purchase, to get the most out of their location and sky conditions. It may be impossible to see deep sky objects from an observing site, so one must concentrate on the brighter solar system objects like the Sun, Moon and planets. Radio astronomy is the same, and it is usually possible to find a radio astronomy project that can be used, even if it means travelling to a more remote site occasionally, as would be done to track down a difficult optical object like the Horsehead Nebula.

Unfortunately with radio astronomy it's not really possible to know how good or bad their location is until they try a project. If it is found there is a problem with interference, it can be treated in the same way as any other problem, sometimes it can be overcome and sometimes it can't. Whatever the location, some useful radio astronomy should be possible.

Here is a short summary of the advantages and disadvantages of each of the four projects covered within the book.

9.1 The INSPIRE Receiver

The INSPIRE receiver and it's antenna takes up very little room, the receiver is battery operated, highly portable, and can fit easily into a backpack. The receiver needs building and is unfortunately highly troubled by mains hum, and one must be prepared to travel to a radio quiet site, possibly in the countryside, to make worthwhile observations. Such efforts will be rewarded, as one will hear some of the strangest sounds ever heard.

9.2 Radio Meteor Detection

Radio meteor detection requires very little space for the receiver and antenna, but one must be prepared to do some research into finding a suitable radio transmitter, and a suitable receiver that is capable of receiving the transmitter's frequency. There can be quite a bit of trial and error with radio meteor detention, and some meteor radiant's may be too high or too low to study effectively. Setting aside all the effort, radio meteor detection is worth it when looking forward to a night's meteor observing and the dreaded clouds roll in, or on a beautifully clear night and longing to be outside to enjoy it, but one has to be at the office first thing for a meeting. In such circumstances simply record the data and process them at a more convenient time.

9.3 The SuperSID Monitor

It really doesn't come much easier than the SuperSID monitor. The SuperSID monitor unit is very small; it takes up no more space than a coffee mug, and comes with its own power supply. All that needs to be done is to fashion an antenna. This takes very basic DIY skills to perform; there are many examples of antenna at the SuperSID website. The down side is that the SuperSID monitor can be prone to interference from household appliances. However, if prepared to keep moving the antenna around until the interference is at an acceptable level and a useable signal is found, the antenna need never be moved again, except to carry out maintenance. The computer software, once up and running, is perfectly happy just doing its own thing and is not at all demanding of attention. It will quite happily keep saving data until one is ready to process it. Once becoming familiar with the SuperSID data it is quite easy to spot X-ray flares within the data, and over the space of a year the changes in the ionosphere overhead can be witnessed.

9.4 The Radio Jove Project

The Radio Jove receiver can be supplied ready-built if wishing to pay a little extra, although much satisfaction can be got from building the receiver, the manual that comes with this receiver is extremely clear and easy to follow. The dual di-pole antennas must be built. The supports and guide ropes for the antennas and a suitable power supply must be found for the finished receiver. The receiver itself takes up very little room, as does the antennas when rolled up neatly for storage, the downside is that quite a large open space is needed to erect the antennas. This isn't a problem if lucky enough to have a large garden or even access to a field, but for many of us this can be a problem. Another downside to the Radio Jove receiver is that if the antennas are used in a noisy environment, it may be impossible to receive the emissions from the planet Jupiter. On the plus side there is still a good chance of picking up emissions from the Sun as these are far stronger.

Whichever project or projects are chosen, it will be sure to provide an interesting accompaniment to the already fascinating hobby of optical astronomy, and one that the dreaded clouds can't spoil!

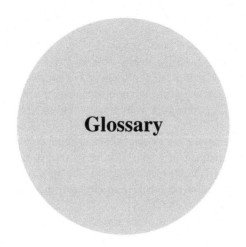

Glossary

AC Alternating current

A.F. Audio frequency. Usually taken to be 20 hertz to 20 kilohertz

Ampere/Amp (A) The unit used to measure electrical current

Amplitude The extent to the oscillation or vibration of a wave

Antenna This is the equivalent of a primary mirror or lens on an optical telescope and can take many forms, and is the collecting area for the incoming signal

Antenna cable Used within to mean coaxial antenna cable

Antenna temperature A measure of the power per unit of bandwidth of a signal received by an antenna

Antenna wire Used within to mean the wire used to make the antennas

Bandwidth The range of frequencies or wavelengths that an antenna is sensitive to

Blackbody A hypothetical body that absorbs all thermal radiation falling upon it and is a perfect emitter of thermal radiation

Blackbody radiation The thermal radiation given off a blackbody. This radiation is in the form of a continuous spectrum

Bridge rectifier A full wave rectifier using four diodes, to turn an alternating current into full wave direct current

Capacitor These act like little reservoirs and store electrons, and come in many types, some of which are polarity sensitive

Conductor A material that offers very little resistance to the flow of electrons through it

Cosmic Microwave Background (CMB) radiation Electromagnetic radiation that seems to emanate from every direction and every point of space, thought to be the faint echo of the "big bang" from the start of the universe

DC Direct current

Decibel Abbreviated to dB. The unit used for expressing transmission gain or loss and relative power levels

Diode A semi-conductor that will only allow current flow in one direction

Di-pole A very popular antenna used by amateurs. Usually cut to half the desired wavelength of the frequency being received

Doppler effect The change in frequency of a wave for an observer, for example the change in pitch of a siren. As the sound wave approach they are compressed and the pitch is higher, and as the siren moves away the pitch drops as the sound waves spread out. Electromagnetic radiation including visible light also exhibits this property

Electric current The movement of electrons through a conductor

Electromagnetic spectrum The whole sequence of electromagnetic wave energy, starting at high energy Gamma rays through visible light to low energy radio waves

Electron A small, naturally occurring particle with a negative charge, that surrounds atoms of matter

Electron volt The unit used to measure the kinetic energy of a particle. $1 \text{ eV} = 1.602 \times 10^{-19}$ Joules

Farad (F) The unit used to measure capacitance

Free electrons Electrons that are not tightly bound to the atoms in a material such as a metal

Frequency The number of repeated wavelengths per second. Frequency is measured in Hertz Hz, but some older books use Cycles per second C/S

Frequency ranges Each frequency has it's own unique properties and uses. They are divided into different frequency bands

Fuse Deliberately designed to be the "weakest link" in an electrical circuit in order to protect the circuit from overload and damage

Harmonic A wave having a frequency that is an integral multiple of the fundamental frequency. For example a wave at twice the fundamental frequency would be called a second harmonic

Henry The unit used to measure inductance

Hertz The unit used to measure of frequency

I.C Integrated circuit (silicon chip), can contain many thousands of transistors and switches, etc. etched onto a wafer of silicon

Io The innermost moon of the planet Jupiter

Io effect Io has the effect of enhancing the radio emissions from the planet Jupiter, by producing a cone of plasma which concentrates the radio waves into a beam

Ionosphere A number of reflective layers in the Earth's upper atmosphere that reflect radio waves and are effected by ultraviolet light, X-rays and other high energy radiation from the Sun and space

Insulator A material that highly resists the flow of electrons through it

Interferometry The use of two or more radio telescopes to observe an object in order to increase the resolving power and increase sensitivity of the telescopes, relative to them being used as individual telescopes

Jansky A measure of radio flux (strength) of a radio source. Named in honour of Karl Jansky. 1 jansky is defined as 1×10^{-26} Wm^{-2} Hz^{-1}

Kelvin A measure of absolute temperature. The point at which no thermal or kinetic energy is present within an atom of material. This is known as absolute zero, or zero degrees Kelvin. To convert degrees Kelvin to degrees Celsius subtract 273. For example 300 degrees Kelvin = 27 degrees Celsius. To convert Celsius to Fahrenheit: Degrees Celsius × 9/5 + 32 = degrees Fahrenheit. Example 27 × 9/5 + 32 = 80.6 Fahrenheit

LED Light emitting diode.

Luminiferous aether A substance that was once thought to exist throughout the universe enabling electromagnetic waves to travel in a vacuum. This was later proven not to exist

Mains power A term used in the United Kingdom to refer to the house hold domestic electric power supply

Magnetosphere A region in space around a planet where the natural magnetic field from the planet deflects the solar wind

Meteor A small particle of debris or meteoroid, in most cases the size of a grain of sand, that burns up in the atmosphere leaving a brief streak of light or meteor trail in the sky

Meteorite A piece of a larger meteor that survived being burnt up by the Earth's atmosphere and has landed on the Earth's surface

Meteoroid A small particle of cometary or asteroidal origin

Micrometeorite A very small meteorite that doesn't burn up in the Earth's atmosphere through the act of friction, but which are quickly slowed down by the atmosphere and gently drifts down through it to the Earth's surface and usually goes unnoticed

Panspermia A theory that life on Earth originated elsewhere and travelled to the Earth by meteorite or comet

Polarisation Whether the incoming electromagnetic waves are vertical or horizontal in its approach to the collecting device, be it a telescopes primary mirror or in radio astronomy an antenna

Polarity In electronics, being negative or positive

Polarity sensitive Some electrical components, like resistors, can be fitted either way round, but some components, like diodes for example, must be fitted in the correct orientation, these are known as polarity sensitive components

Quasars "QUAsi-StellAr Radio Source". A compact object which emits a great deal of energy in a wide range of wavelengths. All quasars have very high red-shifts which indicates that they are some of the most distant and therefore some of the oldest objects in the universe

Quiet Sun The Sun when solar activity is at a minimum

R.F. Radio frequency

Radiant A part of the sky that meteors are seen to originate from during a meteor shower. Meteor showers are usually named after the constellation which contains the radiant

Radio Star The term "radio star" was used quite widely in the early days of radio astronomy. Any radio source in the sky was deemed a radio star because no one really knew what they were. Today the term has been largely forgotten as these radio sources have become better understood. Although the term can still be found in old astronomy books

Receiver The device that amplifies the incoming signal from an antenna and either increases or decreases the frequency of the incoming signal to an audible frequency level

Rectification The act of changing an alternating current to a direct current

Resistor A semi-conductor used in electronics to limit current flow. The unit of resistance is the Ohm and the symbol used to represent this is W

Semi-conductor A material that can, under certain conditions, act as a conductor or an insulator; silicon is such a material

SES Sudden enhancement (of) signal

SI unit(s) The International System of Units. A number of agreed units of measurements that scientists throughout the world use

SID Sudden ionospheric disturbance

Soldering flux Used to clean and prevent oxidization of the components during soldering

Spectrogram A visual graph of a sound or signal with frequency plotted against time

Speed of light 300,000 kilometers per second (186,451 miles per second)

Sporadic meteor A meteor that is not part of a recognized meteor shower and can't be traced back to a known radiant. These can be seen at any time on a clear night and can come from any direction in the sky

Synchrotron radiation Radiation from an accelerating charged particle (usually an electron) in a magnetic field. This can happen at unique wavelengths (lines) by atoms and molecules in space

Thermal radiation Electromagnetic radiation from a body that is often hot. Often characterized by a continuous blackbody spectrum

Toroid collars These are used to help stop interference entering the open end of the antenna cable and also to increase the antenna's performance

Transistor A semi-conductor that has many uses, one of which is as an amplifier

Van Allan belts Two regions of high energy electrons that have been trapped by the Earth's magnetic field. In a torus shape around the planet Earth

Vacuum A volume completely void of matter

Volt (V) The unit used to measure electrical pressure

Voltage The pressure at which electrons are subjected in order to move them through a conductor

Watt (W) The unit used to measure electrical power

Wavelength The physical length of a single wave pattern measured in meters

Yagi A multi-element antenna with high gain. These antennas can be very directional

Zener diode A diode with a specific reverse breakdown voltage

Multiples and Submultiples Used with SI Unit

Multiples	Prefix	Symbol
10^{18}	Exa-	E
10^{15}	Peta-	P
10^{12}	Tera-	T
10^{9}	Giga-	G
10^{6}	Mega-	M
10^{3}	Kilo-	K

Submultiples	Prefix	Symbol
10^{-3}	milli-	m
10^{-6}	micro-	μ
10^{-9}	nano-	n
10^{-12}	pico-	p
10^{-15}	femto-	f
10^{-18}	atto-	a

References

Podcasts

There is an interesting podcast about Karl Jansky at www.365daysofastronomy.org. The date of the podcast is 11th April 2012. This podcast is written and recorded by Dr Christopher Crockett of the US Naval Observatory, as part of Dr Crockett's astronomy word of the week series. The podcast is well thought out and presented. Its well worth tracking down and listening to.

There is a very good podcast available on the internet that is an actual interview with Grote Reber himself. This can be accessed via the Mountain radio website: www.gb.nrao.edu/epo/podcasts.shtml. Reber is interviewed by a member of the Greenbank radio telescope staff. He talks about how he built his radio telescope and the problems that he had to overcome in doing so. He also mentions how everyone thought he was an eccentric and a bit of a crank.

There are seven podcasts relating to the sections of the electromagnetic spectrum and their use in astronomy at www.astronomycast.org. Each podcast is recorded as a discussion of the subject at hand by Fraser Cain and Dr Pamela Gay, and these podcasts can be very entertaining at times and well worth listening to:

130 Radio Astronomy.
131 Submillimeter Astronomy.
132 Infrared Astronomy.
133 Optical Astronomy.
134 Ultraviolet Astronomy.
135 X-ray Astronomy.
136 Gamma ray Astronomy.

There is a short podcast about the SuperSID monitor that can be found at www.365daysofastronomy.org. The date of the podcast is 21st April 2010. It's written and recorded by Jim Stratigos, and he's made a cracking job of condensing the subject into a 15 minute long podcast. He discusses with one of the SuperSID team how the SuperSID monitor came into being, and how students have found it a great introduction to the subject of studying the Sun. The podcast is a great source of information and well worth the effort to track it down.

There is a podcast about the Radio Jove project and receiver, it is available at www.365dayofastronomy.org. The date of the podcast is 29th April 2011. This podcast gives an abridged version of the Chap. 7 "The NASA Radio Jove project".

At the INSPIRE project website http://theinspireproject.org/ there were two short podcasts about the INSPIRE receiver. One was of a recording from a radio interview and the other was a very short introduction to the receiver. Unfortunately the last time the website was visited the radio interview podcast couldn't be found. This may have been temporarily removed. If one has the time they are both worth tracking down, if only to hear the INSPIRE team trying to mimic the sounds picked up by the receiver during the radio interview.

Jodrell Bank radio observatory in the United Kingdom produces a twice-monthly podcast which they have nicknamed "Jodcast". Within the Jodcasts are all the up-to-the-minute advances in the field of radio astronomy, also what is happening within the night sky over the next month and what astronomy events to watch out for.

The website Cheap Astronomy www.cheapastro.com produces a podcast every Wednesday and the host of these podcasts is Steve Nerlich. All of Steve's podcasts are entertaining, and all of them are done in Steve's unique style. So beware if listening to Cheap Astronomy podcasts wearing headphones in a public place, as his podcasts can make one laugh out loud. Look through the list of podcasts and find podcasts about the square kilometer array and other podcasts of interest to the radio astronomer.

The ordinary guy from the brains matter website at http://www.brainsmatter.com/ has amongst his list of podcasts, which cover a wide range of subjects, several on the subject of optical astronomy and one or two on radio astronomy. These are highly recommended and worth tracking down and listening to.

Books

Radio Nature. By Renato Romero.
An excellent book for anyone who wishes to pursue the hobby of very low frequency observations. Easy reading with very little mathematics. The INSPIRE project is briefly covered, as are low frequency radio emissions from the Sun. It has some good illustrations and images, including one image in particular: a slightly strange image of a submarine's antenna.

References

Whistlers and Related Ionospheric Phenomena. By Robert A. Helliwell.
This book goes into great detail about the subject and can be a little heavy going at times. There is also quite a bit of mathematics involved.
  Nevertheless, it is a good book that would be of great value if one is thinking of going into more advanced VLF studies where a greater knowledge of the subject of ionospheric phenomena is required.

Listening to Jupiter: A Guide for Amateur Radio Astronomers, Second Edition. By Richard S. Flagg.
This book is very highly recommended. It is written in a friendly and easy to read manner and there are several funny anecdotes, such as the one about how flames shot out of a project he had recently finished building, and one about the night he came across an alligator. If anyone has the remotest interest in trying to observe the planet Jupiter and the Sun's radio emissions, buy this book.

The ARRL Antenna Book.
This book covers all aspects of antennas and is handy to have lying around to refer to. It is a must have if one decides to start designing and building their own antennas, or just wishes to have a greater understanding of how antennas work. The book also contains information on the suppliers of antennas and other connected items.

Radio Astronomy. By F. Graham Smith.
This book is a bit of a dinosaur now, as it was written in 1960, in the "good old days" before computers, where vacuum tubes (valves) reigned supreme. Its value is more historic now. Nevertheless, it is still a good read if one can find a copy. Note that cycles per second are used instead of Hertz Hz to denote frequency, and that the Andromeda galaxy is referred to as a nebula. The book is a snapshot from the beginning of radio astronomy.

Amateur Radio Astronomy. By John Fielding.
An interesting book that covers each aspect of a radio telescope separately. This book would be useful for someone who has some radio and electronics skills rather than an absolute beginner, but due to advances in technology some of the equipment is dated.

Radio Astronomy Projects. By William Lonc.
This book contains a large number of radio astronomy projects, HOWEVER these projects are not for the faint hearted and require some serious kit and a deep pocket. It is recommended one has some experience and knowledge of the subject before embarking on some of these more advanced projects.

The Invention that changed the world the story of RADAR from war to peace. By Robert Buderi.
Nearly 600 pages in length, covering all aspects of RADAR from its beginnings. There are sections within the book that tell how RADAR has been used in astronomy. A good read if one wants to know more about the history of RADAR and RADAR astronomy.

The Art of Soldering. By R. Brewster.
This book covers all aspects of the art of soldering. It would be a good investment as it is aimed at the beginner and covers the different types of soldering irons and solder. It also explains how to make a good soldered joint and how to avoid some of the common pitfalls that one may come across when first starting.

Getting the most from your multimeter. By R.A Penfold.
If thinking about getting a multimeter, or already have one and not sure what to do with it other than checking to see if a battery requires changing, this book will be useful. This book does exactly what the title says, how to get the most from a multimeter. It explains what kind of tests can be carried out using a multimeter and how to connect the multimeter to an electrical circuit in order to carry out such tests as measure voltage, current and resistance. It is written in a very user-friendly way and is aimed at the beginner.

How to use oscilloscopes and other test equipment. By R.A Penfold.
This book is the follow-up to the book "Getting the most from your multimeter". This book covers more advanced test equipment, such as oscilloscopes, and is aimed more towards the intermediate person rather than the beginner. If thinking of purchasing the more advanced test equipment the book would prove useful in knowing what to look for in a piece of equipment.

Secrets of the Sun. By Ronald Giovaella.
Mainly for the optical astronomer, but the book gives some excellent examples of actual images taken of the Sun. The book is a few years old now, but it still contains some very useful information. It covers in detail space weather, sunspots, flares and prominences. The solar cycle and the Sun's magnetic field are also covered, the motions of the gases within the Sun's atmosphere are also discussed.

Solar Observing Techniques. By Chris Kitchin.
This book covers all aspects of observing the Sun optically and in different wavelengths, such as hydrogen alpha. One small criticism is the use of foam filter mounts used on the telescopes pictured within the book, as these can be dangerous, but there is a warning given about these within the book itself. Also a short chapter of 5 pages covering radio observations made of the Sun.

Sun. By Pam Spence.
Aimed at the optical astronomer this is a handy little book that covers all visual parts of solar astronomy. Ideal for someone thinking of taking up solar astronomy especially on a budget, or if someone has an old 60 millimeters (2.36 inch) refractor laying about and is wondering what to do with it other that using it to look at the Moon. The solar cycle, sunspot classification, and the Sun's magnetic field are also covered within the book.

Meteorites. By Robert Hutchison and Andrew Graham.
A short book of 60 pages, full of information about the different types of meteorites that have been found through-out the world. It contains some interesting colored images of the crystalline structures of meteorites, and describes how observing

them with polarized light can show up different chemical elements within the thin cross-sections of meteorites. A good starting point for the study of meteorites. The book is written in a non-technical manner.

Field Guide to Meteors and Meteorites. By O. Richard Norton and Lawrence Chitwood.
This book has been very well thought out and put together. It covers meteors and meteorites in great detail and discusses the different types of meteorites, both stone and metallic types and the physical properties of each. It is written in a highly readable way and anyone remotely interested in these "visitors" from outer space would do well to get a copy.

Software

An excellent piece of software for use by the budding radio astronomer is the program "Radio Eyes". This is available from Radio Sky Publishing. The best description of this program is a radio version of an optical planetarium program, having interesting radio objects instead of interesting optical objects. Please see screen-shot in Fig. A.1.

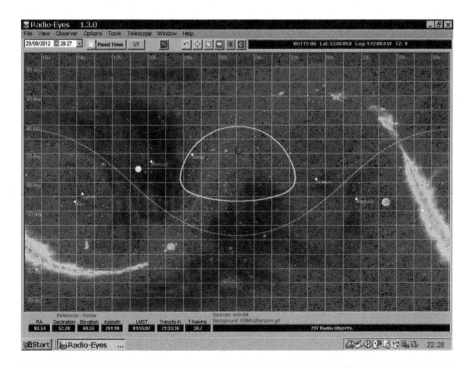

Fig. A.1 Screenshot of the program "Radio Eyes"

This program has a clever feature, where one can set the beam pattern of a radio telescope and this will then be projected onto the sky, by clicking onto one of the boxes in the top right a list of objects will be listed that are within the telescopes beam pattern. It can be advantageous to know if there are strong radio sources in the same part of the sky as the planet Jupiter, especially if trying to receive Jupiter's radio emissions. The background of the program is colored to represent the different levels of radio noise from that particular part of the sky. At the default settings a light green color indicates areas of high radio noise and the colors purple and black show the quieter parts of the sky. Each radio source is marked on the sky with a little pink circle. Click on one of these circles, an information window pops up showing the name of the radio source and relevant information pertaining to the radio source, such as its position, the frequency that it can be received and the strength of the radio flux measured in Janskys.

If choosing to have pulsars shown on the map, these show up as small yellow circles when the program is on default settings. These can be clicked on to reveal further information, such as its position, frequency, spectral class and rotational speed.

Another good feature of this program is that the view can be changed to show a dome view. This looks just like the views shown within the chapter on the Radio Jove project and the Radio Jupiter pro program. Having both the east west flat window and the dome view window open at the same time one gets a three dimensional view of the sky. Also, as the computer cursor is moved across either the dome view window or the east west view window, a corresponding cursor will appear within the other window to give the exact position upon the sky. This positioning can be further refined by the program's zoom feature. This is very handy when trying to find an object in a particularly crowded part of the sky. Just click the mouse button and draw a rectangular marquee around the part of the sky and the program will automatically zoom in to this point.

It can also show the ecliptic, constellation patterns, and solar system objects such as the planets, Moon, and Sun on the screen. Things like the changing phases of the Moon are not shown, but the Moon's position relative to other objects is.

There is also a search feature, enter the name of an object and the program will search through its database, find the radio source, and point to it. A word of caution, unlike other planetarium programs that one has used, where if the spelling is incorrect other programs will give the closest match to the name that is typed in, the Radio Eyes program will only accept the correct spelling for an object. This should be seen as only a very minor criticism of an otherwise excellent program.

The program has a lot of other interesting features, one of which is the capability to control a telescope mount in both altitude and azimuth. As yet this feature has not been used and no further comment about how well it performs this task can be made.

The software can be downloaded from http://radiosky.com. The software comes with a 30-day free trial period and after this date the license must be purchased. This is quite easy to do over the internet. Once the payment for the license has been received, an email will be sent with a code; simply type this code into the program and enjoy. Any updates and improvements to the program are free to download and install, therefore be sure to always have the up-to-date version.

Another useful website for software, is www.weaksignals.com. This website has some useful software that may be of assistance to the more advanced radio astronomer enthusiast.

Websites

Jodrell Bank United Kingdom http://www.jb.man.ac.uk
Greenbank radio telescope United States http://science.nrao.edu/facilites/gbt/
Parkes radio telescope Australia http://www.parkes.atnf.csiro.au
Max Planck observatory Germany http://www.mpifr-bonn.mpg.de
Arecibo Observatory Puerto Rico http://www.naic.edu
Royal Air Force Air Defence RADAR Museum www.radarmuseum.co.uk
Radio Jove Project http://radiojove.gsfc.nasa.gov/. Download an order form from this site to get a Radio Jove Receiver kit.
Radio Sky Publishing's http://radiosky.com. From the website download software such as Radio Jupiter pro, Radio-Skypipe, and Radio Eyes, plus books such as Listening to Jupiter. This site is highly recommended and well worth a visit.
A very useful site for the study of very low frequency electromagnetic waves is http://www.vlf.it/. This site is highly recommended if interested in VLF signals. The book "Radio Nature" can be purchased from this website.
A useful site for radio meteor detection http://home.deds.nl~knol
The INSPIRE project http://theinspireproject.org/. Order an INSPIRE receiver kit from this website.
Society of Amateur Radio Astronomers (SARA) www.radio-astronomy.org
Stanford University (SuperSID monitor) http://solar-center.stanford.edu/SID/. Order a SuperSID monitor from this website.
Society for Popular Astronomy (SPA) www.popastro.com
British Astronomical Association (BAA) http://britastro.org/
Search for Extra-Terrestrial Intelligence SETI league http://www.setileague.org
Solar and Heliospheric Observatory (SOHO) http://sohowww.nascom.nasa.gov/
This is a good website to visit to keep up-to-date with solar activity such as sunspots and flares.
United Kingdom Radio Astronomy Association (UKRAA) http://www.ukraa.com/
Radio Society of Great Britain (RSGB) http://www.rsgb.org/. This website has a number of books on the construction of antennas that may be of interest to the radio astronomer.
The National Association for Amateur Radio (ARRL) http://www.arrl.org.
From this website one can order the ARRL antenna book.

From this website www.astrosurf.com/luxorion/audiofiles.htm all manner of sounds from space can be heard. These include pulsars, lightning discharges on Jupiter and Saturn and the sound made by the aurora at their poles. There are also the recordings made by the two Voyager space probes as they flew close to the magnetosphere of Jupiter.

Index

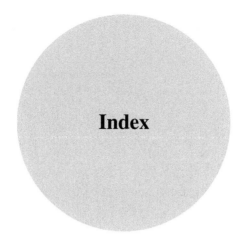

A
Absolute zero, 25, 35, 47, 158
Amplitude, 37, 86, 127, 198
Antenna, 1, 2, 4–6, 8, 10, 12, 14, 16, 18, 21, 24–26, 38, 39, 42, 45, 47, 54, 58, 61, 63, 73, 81–83, 85, 86, 88–96, 98–102, 105, 106, 110, 111, 114–118, 121, 123, 124, 127, 133, 135–137, 140, 144–154, 156–162, 164, 166, 172, 175, 177–180, 184, 187–189, 192, 193, 197–201
Antenna temperature, 158, 159, 198
Arecibo radio telescope, 23, 28, 29
Atmospheres of planets, 41
Audacity, 182, 183
Audio frequency, 44–45, 77, 142, 198

B
Bandwidth, 30, 158, 172, 198
Bell, J., 26, 28
Bernard, L., 10, 11, 15, 18
Blackbody radiation, 198
BNC connector, 81, 93, 94, 98, 179
Burke, B.F., 16, 25, 164

C
Calibration, 77, 154, 158–160
Capacitors, 69, 71–72, 75, 110–113, 136–139, 143, 198
Cassiopeia A., 13, 22

Cavity magnetron, 8, 9
Chorus, 125
Cosmic background radiation (CMB), 24–26, 198
Crab pulsar, 48
Cycles per second, 38, 193, 199

D
De-soldering pump, 64, 65
De-soldering wick, 65
Diodes, 69, 72–73, 75, 109, 113, 138, 198–201
Di-pole antenna, 137, 144–148, 150–152, 160, 178, 179, 189
D-layer ionosphere, 52, 53, 86
Doppler effect, 199
Doppler shift, 172, 176, 181
Dynamo effect, 55, 60, 124

E
Earths core, 27, 36, 55, 56, 60, 124
Edison, T.A., 1, 2, 160, 161
E-layer ionosphere, 52, 53
Electromagnetic spectrum, 39–42, 47, 191, 199
Electron(s), 14, 27, 33–36, 38, 46, 47, 56, 71, 72, 113, 122, 198, 199, 201
Electron volt, 199
Ewen, H.J., 14

Exosphere, 50
Extremely high frequency, 43
Extremely low frequency, 44, 109

F
Farad, 199
F-connector, 144, 148
F-layers ionosphere, 52, 53
Franklin, K., 16, 164
Free electrons, 34, 199
Frequency, 3, 6, 9, 10, 12, 16, 17, 19, 20, 30, 31, 37–38, 42–45, 47, 51, 53, 54, 57, 73, 77, 78, 86, 99–101, 109, 118–120, 122–124, 126, 128–130, 133, 135, 142, 160, 161, 164, 172, 173, 175–177, 179, 181, 183, 188, 192, 193, 196, 197, 199–201

G
Gamma rays, 39, 40, 50, 191, 199

H
Harmonics, 120, 129, 199
Headphones, 44, 77–79, 98, 110, 118, 119, 121, 136, 142, 143, 159, 176, 180, 192
Heat shunts, 65, 74
Henry, 199
Hertz, 1, 38, 39, 44, 45, 77, 109, 118–120, 128, 130, 181, 193, 198, 199
Hey, J.S., 8–10, 13, 161, 171
High frequency, 3, 43, 73
Horn antenna, 24, 25, 47

I
Inductor, 74–75, 113, 137, 142, 143
Infrared, 41–42, 47, 50, 191
INSPIRE project, 3, 109–133, 180, 192, 197
INSPIRE receiver, 53, 54, 68, 109–120, 122, 126, 127, 133, 181, 188, 192, 197
Integrated circuits (ICs), 69, 75–77, 112, 114, 137, 139, 199
Interference, 3–5, 7, 10, 16, 25, 31, 45, 54, 78, 93–95, 99, 101–103, 118–121, 146, 150, 159, 160, 164, 167, 181, 183, 187–189, 201
Interferometer, 12, 13, 21
Interferometry, 12, 21–22, 28, 199
Io, 58, 60, 155, 156, 165, 166, 199
Io effect, 60, 165, 199
Ionisation, 34, 52–54, 86, 87, 100, 101, 103, 165, 166, 171, 172, 175, 181

Ionosphere, 3, 9, 10, 17, 18, 41, 44, 50–54, 56, 57, 81, 83, 85–87, 96, 100–103, 109, 123, 124, 161, 165, 166, 189, 199

J
Jansky the unit of, 5
Jodrell Bank, 10, 12, 14, 15, 17, 18, 21, 192, 197
Jupiter, 1, 16, 28, 34, 41, 47, 48, 53–55, 57–60, 127, 128, 135–137, 144, 148, 150, 153–157, 160, 161, 164–167, 189, 193, 196, 197, 199

K
Karl, J., 3–5, 24, 156, 160, 191, 200
Kelvin, 39, 46, 158–160, 175, 200

L
Laika, 17, 18, 51
Lightning, 34, 59, 62, 88, 94, 102, 122–124, 160, 164, 197
Little green men (LGM), 26
Lodge, O.J., 2, 161
Low frequency, 3, 42–44, 53, 54, 57, 109, 133, 192, 197
Luminiferous aether, 2, 36, 200

M
Magnetic field of Jupiter, 47, 48, 55, 57–60, 165
Magnetic fields terrestrial planets, 55
Magnetosphere of Jupiter, 197
Magnetosphere of the earth, 55–57, 59, 60, 125
Mars, 46, 49, 55, 84, 158, 170, 171
Medium frequency, 43
Mercury, 23–24, 45, 49, 55, 58, 59
Mesosphere, 50
Multimeter, 66–68, 71, 93, 141, 142, 194

N
National Aeronautics and Space Administration (NASA), 11, 21, 50, 59, 83, 88, 106, 109–133, 135–167, 170, 192
Noise, 3, 12, 25, 26, 30, 45, 47, 71, 77, 97, 99, 118–130, 141, 142, 155, 156, 159, 160, 162, 165, 175, 181, 183, 196

Index

O
Ohm's law, 35
Orientation of components, 69

P
Panspermia, 171, 200
Penzias, A., 24, 25, 47
Phasing cables Radio Jove, 144
Polarization of electromagnetic waves, 39
Power supplies, 141
Pulsars, 26–28, 40, 48, 160, 196, 197
Purcell, E.M., 14

Q
Quasar, 21–22, 48, 200
Quiet Sun, 162, 200

R
Radio eyes, 195–197
Radio Jove project, 135–167, 189, 192, 196, 197
Radio Jove receiver, 2, 8, 38, 47, 53, 54, 57, 60, 68, 71, 81, 110, 111, 126, 127, 135–145, 147, 150, 154, 155, 157–160, 162, 164–167, 179, 189, 197
Radio Jupiter Pro, 136, 150, 154–156, 165, 166, 196, 197
Radio star, 13, 201
Radio waves, 1–3, 5, 11, 13, 15, 16, 19, 24, 45, 50, 51, 53, 54, 59, 60, 123, 144, 156, 165, 172, 181, 199
Reber, G., 5–8, 13, 191
Rectification, 72, 141, 201
Resistors, 69–75, 109–113, 115, 136–140, 143, 200, 201

S
Safety, 8, 61–62, 94
Sagittarius A, 5, 7, 24, 31, 156, 160
Scintillation, 54–55
Sferics, 122, 123, 127
SI unit(s), 201, 202
Society of Amateur Radio Astronomers (SARA), 82, 83, 106, 107, 110, 184, 197
Solar radio emission classification, 167
Soldering irons, 46, 61–65, 74, 147, 187, 194
Solder types of, 63
Sound cards, 95–99, 129, 154, 176, 180
Space weather, 81, 84–85, 105, 106, 194
Spectran, 129–133, 180, 183
Spectrum Lab, 131–133, 180, 183
Stanford solar center, 81–107
Stratosphere, 50
Sub-hertz frequency, 44, 109
Submillimeter, 42, 43, 191
Sudden Enhancement of Signals (SES's), 81, 86, 87, 96, 103, 201
Sudden Ionospheric Disturbances (SID's), 57, 86, 87, 96, 102, 103, 197, 201
Sunrise-sunset effect, 102
Super high frequency, 43
Super low frequency, 44, 109
SuperSID monitor, 52–54, 81–107, 154, 179, 188–189, 192, 197
Synchrotron radiation, 46–48, 201

T
Thermal radiation, 46, 47, 162, 198, 201
Thermosphere, 50
THz frequency, 42
Toroid collars, 93, 144, 147, 201
Transistors, 6, 65, 69, 73–76, 137, 199, 201
Troposphere, 50
Tweeks, 123, 124, 127

U
Ultra high frequency, 43
Ultra low frequency, 44, 109
Ultraviolet, 16, 34, 40, 41, 50–52, 84, 149, 191, 199

V
Van Allan belts, 201
Venus, 16, 17, 28, 34, 41, 45, 49, 55
Very high frequency, 43
Very low frequency, 3, 42–44, 53, 54, 109, 133, 192, 197
Visible light, 11, 40–42, 50, 59, 199
Voltage, 35, 67, 71–74, 85, 105, 118, 120, 141, 142, 194, 201

W
Watt, 10, 62, 63, 70, 147, 201
Wavelength, 8, 11–16, 22, 23, 25, 32, 37–47, 50, 51, 53, 59, 85, 86, 123, 144–146, 156, 162, 178, 179, 194, 198–201
Whistlers, 54, 57, 123–126, 129, 193

Wilson, R., 24, 25, 47
WOW signal, 28–32

X
X-ray(s), 40, 42, 43, 50–52, 56, 81, 83–88, 103–105, 162, 189, 191, 199
X-ray flare classification, 105

Y
Yagi antenna, 10, 179

Z
Zener diode, 72, 73, 113, 138, 201

Printed by Printforce, the Netherlands